The Institute of Biology's
Studies in Biology no. 51

The Ecology of
Small Mammals

M. J. Delany

D.Sc.

Professor of Environmental Science,
University of Bradford

Edward Arnold

© M. J. Delany 1974

First published 1974
by Edward Arnold (Publishers) Limited,
25 Hill Street, London W1X 8LL
Reprinted 1976

Boards edition ISBN: 0 7131 2473 3
Paper edition   ISBN: 0 7131 2474 1

Printed in Great Britain by
The Camelot Press Ltd, Southampton

General Preface to the Series

It is no longer possible for one textbook to cover the whole field of Biology and to remain sufficiently up-to-date. At the same time teachers and students at school, college or university need to keep abreast of recent trends and know where the most significant developments are taking place.

To meet the need for this progressive approach the Institute of Biology has for some years sponsored this series of booklets dealing with subjects specially selected by a panel of editors. The enthusiastic acceptance of the series by teachers and students at school, college and university shows the usefulness of the books in providing a clear and up-to-date coverage of topics, particularly in areas of research and changing views.

Among features of the series are the attention given to methods, the inclusion of a selected list of books for further reading and, wherever possible, suggestions for practical work.

Reader's comments will be welcomed by the author or the Education Officer of the Institute.

1974
                                    The Institute of Biology,
                                    41 Queens Gate,
                                    London, SW7 5HU

Preface

The free living small rodents and insectivores are groups of mammals of interest to many professional and amateur biologists. Not only are they interesting little animals in their own right but by virtue of their abundance and wide range of ecological adaptations they are also important components of almost every existing terrestrial ecosystem. Their relationship to man witnesses its greatest impact when they become serious pests of agriculture and forestry and transmission agents of disease.

Their small size, relatively short life cycle and high reproductive capacity make them ideal subjects for the study of mammal ecology. During the past two decades research workers in both temperate and tropical regions have taken advantage of this suitability of these animals to detailed investigation and have, in consequence, accumulated a great deal of information on them. The purpose of this volume is not to provide a comprehensive review of their results but to focus attention on the major lines of research that have been developed, describe the principal methods that have been adopted and draw attention to the broader ecological issues that these studies highlight. Advantage has been taken of the cosmopolitan distribution of small mammals to refer to a diversity of habitats and climates so as to place this work in its broadest possible context.

Southampton 1974   M. J. D.

Contents

1 What is a 'Small Mammal'?

Within the ecological literature the term 'small mammal' not infrequently appears—without quotation marks—and with the general assumption that the reader will be acquainted with the term. In fact, it usually has a very special meaning and certainly does not embrace all kinds of small mammals. If it did some bats would undoubtedly be included, which is generally not intended. In the present study also 'small mammal' has a fairly specific meaning. It is basically intended to include the free-living small rodents and insectivores.

The lower size limit is set by the animals themselves with the Etruscan shrew (*Crocidura etruscus*), the smallest known mammal, weighing as little as 2 g. The upper limit is more difficult to define as there is a gradation in size to the very large species. A useful and arbitrary measure is to include animals up to about 120 g weight; this approximates to the largest size that can be regularly caught in a commercially produced break-back rat trap. There is a considerable range of species within these size limits, including many shrews and moles among the insectivores and most rats, mice, lemmings, gerbils, jerboas, dormice and some of the smaller squirrels among the rodents. These are the small non-flying mammals found on and under the ground surface and amongst the vegetation in natural and semi-natural habitats. In the United Kingdom we have seven small mammals that are widespread and common. (Others do occur but they are either of restricted distribution or are nowhere very abundant.) Three of these are rodents—the wood mouse (*Apodemus sylvaticus*) with its long tail and large eyes and the bank and field voles (*Clethrionomys glareolus* and *Microtus agrestis*) with their blunt faces, short tails, small eyes and in the former a red-brown fur contrasting with the golden-brown of the latter. The insectivores include the pygmy (*Sorex minutus*), common (*S. araneus*) and water (*Neomys fodiens*) shrews and the mole (*Talpa europa*). Identification of small mammals here and elsewhere in the world is usually not difficult. The main problems arise in places where they have been little studied both ecologically and systematically. A list of useful guides to identification is given in the Appendix.

As a result of the nature of their habitat the study of the ecology of house rats and mice closely associated with man has been along rather different lines to other mammals occupying more natural situations. Furthermore, a great deal of research has been directed to the control and ultimate eradication of these pests. Whilst this is obviously a very important and interesting area of study, space does not permit detailed consideration of the ecology of these animals in this account.

2 Field Methods

2.1 Choice of method

Even though small rodents and insectivores are very numerous their secretive habits and small size result in their being seldom seen in the wild. This makes direct field observation impracticable. The obvious and most suitable alternative is to trap them and obtain from the catch the required information. The precise methods adopted will depend on the question being posed and it is therefore very important to clearly define the problem at the outset. Unfortunately, mammals are not the easiest of animals to study in this indirect way as they often show individual variation in behaviour to traps as well as behavioural changes during the course of the life history. For these reasons it is important to incorporate precautions that minimize bias. Finally, in devising a serious study, care must be taken to balance the maximum return of useful results against practicality in terms of available time. Fieldwork can be extremely time-consuming and what may seem a most suitable study from the armchair can prove a totally unrealistic assignment in the field.

2.2 Traps

One of the most popular and widely used traps in this country is the Longworth small mammal trap (CHITTY and KEMPSON, 1949). This aluminium trap is made of two sections, a tunnel and a nest box (Fig. 2–1b). The animal enters the tunnel and steps on a treadle sensitive to 1.5 g at the end remote from the door which then closes and locks. The nest-box containing bedding and food ensures that the mouse or vole is kept warm and dry until the trap is examined. The tunnel entrance measures 50 × 62 mm which effectively excludes rodents above 60 g weight. A small metal sheet containing a 1 cm diameter hole can be inserted in the tunnel between the door and the treadle. This will exclude all but the youngest mice and voles but permit the entry of the shrews encountered in Europe (but obviously not, for example, the larger African *Crocidura* which may weigh up to 60 g.). This trap is efficient, light (250 g) and weatherproof. Its main disadvantage is its bulk, as those of us who have transported a case of fifty through the Hebrides can testify!

A slightly different live trap is the Sherman (Fig. 2–1a). This consists of a tunnel with a door at one end and a sensitive platform at the other. As the mouse walks on this a spring is released which closes the door. They can be obtained made of 18.12 kg tin or 0.20 gauge aluminium and are manufactured in two sizes 50 × 62 × 165 mm and 76 × 89 × 229 mm. One of their very real advantages is that they can be folded to a thickness of about 15 mm. The principal shortcoming is the lack of room for the provision of bedding and a consequent high mortality of the catch if the traps are not very frequently examined in cold weather. But this is an excellent trap for use in warm climates.

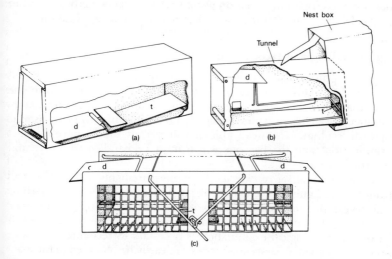

Fig. 2–1   Three types of live trap **(a)** Sherman, **(b)** Longworth, **(c)** Havahart; t, sensitive treadle or platform, d, closing door.

The Havahart trap (Fig. 2–1c) is basically a galvanized steel cage with a door at each end and a sensitive treadle between. Weight on the treadle releases the struts holding the door open, with reopening prevented by a simple locking device. These traps are manufactured in several sizes; the smallest size (o), having a door opening of 76 × 76 mm, is most suitable for small mammals. The catch can be clearly seen without opening the trap but the lack of nest-box, bulk and an exceptionally sensitive tripping apparatus that can be released by heavy rainfall, impose limitations on their use.

If it is not live animals that are required various suitable commercially produced break-back traps are available. These are made with both wood and metal bases and generally supplied in 'rat' and 'mouse' sizes. Care has to be taken to ensure that the sensitive portion carrying the bait, whether this be a pair of prongs or a small platform, will spring the trap with the slightest pressure. A rather more specialized break-back mouse trap which is designed not to smash the skull of the animal is the Museum Special. It is frequently desirable to retain these traps by tying them to pieces of nearby vegetation with string, otherwise they can be moved by an animal that is caught and not killed, as well as by predators feeding on the dead catch. With snap traps generally the springs can, with protracted use, experience fatigue and should be checked regularly.

All these traps are primarily designed for use on the ground but there is no reason, apart from the difficulty in getting them satisfactorily attached, why they should not be placed in vegetation. DELANY (1971) has obtained useful information using snap traps above ground level in tropical forest, but this technique has not had wide application.

2.3   Handling and recording the catch

The captured animal is an important source of very useful information. This may not at first sight be very obvious from the examination of a single specimen but the accumulated data on a large number of catches provides extremely useful insight into many aspects of their ecology. The dead animal will permit much more detailed examination than the live one which must be handled with care and returned to the field as quickly as possible.

A conventional set of external body measurements can be obtained from the dead animal (for details see SOUTHERN, 1964), whilst dissection will reveal the number of embryos present (including the number resorbing), the number of recent placental scars and the occurrence of lactation. Histological examination is necessary to recognize early and previous pregnancy. Descended testes can easily be observed in the male and their activity measured through the microscopic examination of sperm from the vesiculae seminales (COETZEE, 1965). This reproductive data is invaluable in reconstructing the breeding pattern and recruitment rate of new animals into the population. The demographic picture becomes much more complete if the age of each animal is known. Unfortunately, there is no simple and reliable method of ageing available. Injection of live bank voles with the vital calciphil dye sodium alizarine sulphonate (Red S) results in the deposition of a red band dye at the time of injection in the continually growing root of the first lower molar (LOWE, 1971). Repeated injection produces more bands so that it is ultimately possible to measure the growth rate of the root and

Fig. 2–2   Ageing by examination of molar teeth. (a) First left mandibular molar of the bank vole; the dye rings are laid down at the time of each injection and their distances measured from the crutch with a microscope eye-piece scale. Months represent trapping and marking periods and figures distance from crutch. Note how the root does not have a uniform growth rate (from LOWE 1971; courtesy *J. Anim. Ecol.*), (b) Three stages of dentine exposure in the upper molars of the harsh-furred mouse. (From DELANY, 1971; courtesy Zoological Society of London)

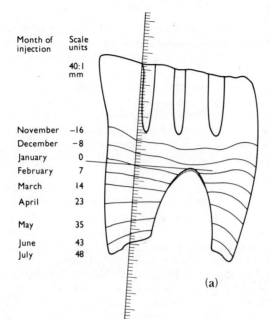

Month of injection	Scale units
	40:1 mm
November	−16
December	−8
January	0
February	7
March	14
April	23
May	35
June	43
July	48

(a)

Dentine

Dentine

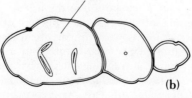

Dentine

(b)

1mm

calculate from this the age of the animal (Fig. 2–2a). Indices of age can be obtained from measurements of tooth wear. This method is most appropriately applied to large samples collected over a twelve month period. Examples include measurement of the height of the lower canine of shrews (CROWCROFT, 1956), reduction in cusp number of the upper molar tooth row in *Apodemus* (DELANY and DAVIS, 1960) and the increasing dentine exposure (Fig. 2–2b) in the harsh-furred mouse (DELANY, 1971). But it should be remembered that rates of wear vary according to the type of food being consumed and may consequently vary between localities and at different times of the year at one place.

Live animals provide only a limited amount of direct observational information, e.g. sex, weight, presence of descended testes, perforate vagina, etc., with their principal ecological contribution usually resulting from the recapture of diagnostically marked individuals. Several marking methods are available. They include toe clipping or the ectomizing of the terminal joint (not just the nails) of two digits. This is best done under local anaesthetic, e.g. chloroethane, with a sharp pair of scissors. Alternatively, numbered rings (LINN and SHILLITO, 1960) can be carefully fitted to the hind foot so as to prevent loss and constriction, but they must be sufficiently strong to resist the gnawing teeth. Numbered ears tags (STODDART, 1970) are available. However, their relatively large size probably limits their use to a limited number of species. Dyes are generally not very satisfactory, except in the short term, as they disappear with moult and possibly washing.

Radioisotopes have been used in a variety of ways and the animals or their faeces then tracked with Geiger-Muller detectors. GENTRY *et al.* (1971) introduced ^{59}Fe, ^{65}Zn and ^{131}I into four species by including these substances in peanut butter on which they fed, GODFREY (1955, 1957) placed tail rings containing ^{60}Co on the European mole and MILLER (1957) injected sodium orthophosphate in buffered solution labelled with ^{32}P into the peritoneal cavity of *Microtus*. As it is not possible to distinguish between different animals unless more than one isotope is applied, it may be desirable to have only one marked individual at a time within a particular area.

Whenever information is obtained in the field it should be recorded immediately in a notebook (not on loose sheets that are easily lost) and not left stored, even for a few hours, in the usually slightly unreliable memory. The keeping of accurate and full records is every bit as important as undertaking the experiments themselves. How they are kept is largely a matter of personal choice, but a useful and simple method of summarizing a trapping survey has been proposed by PETRUSEWICZ and ANDRZEJEWSKI (1962) in their Calendar of Catches (Fig. 2–3). This provides an easy reference to the population composition at a particular time.

Week	1	2	3	4	5	6	7	8	9
Date	| August |				| September |			
	1	8	15	22	29	5	12	19	26
1 ♂	●——	——	●						
1 ♀	●—	●	——	●	——	●			
2 ♀	●								
3 ♀	●								
2 ♂			●——	——	●	——	●		
3 ♂		●—	—●						
4 ♀					●——	——	—●	—●	
5 ♂					●				
6 ♂						●—●	—●		
7 ♂							●		
Caught	4	3	2	3	2	3	3	1	0
Present ♂	1	3	3	1	3	2	2	0	0
Present ♀	3	1	1	2	2	2	1	1	0
Present Total	4	4	4	3	5	4	3	1	0
New	?	2	0	1	2	0	1	0	0
Disappeared	2	0	2	0	1	2	2	1	
One appearance	2	0	0	0	1	0	1	0	

Disappeared = last appearance on this date

Fig. 2–3 Calendar of catches. (After PETRUSEWICZ and ANDRZEJEWSKI, 1962)

2.4 Home range

The home range of an animal can be simply described by JEWELL'S (1966) restatement of BURT'S (1943) definition, viz. *home range is the area over which an animal normally travels in pursuit of its routine activities.* It is not necessarily fixed for the whole of the animal's life (JEWELL, 1966; BROWN, 1966; SHILLITO, 1963) and can vary considerably in relation to population density (STICKEL, 1960). Distinction should be made with *territory* which is essentially, *any defended area.*

Home range in small mammals is usually measured by mark, release and recapture of individuals in live traps. It can be accompanied, if animals are marked by toe-clipping, by tracking. This involves placing plates containing a thin surface, e.g. suspension of fine talc in a water-repellant medium in open-ended containers. The pressure of the foot then leaves an impression on the plate (BROWN, 1966). Tracking helps to minimize the limitations such as trap addiction and holding for long periods, imposed by using traps alone.

Home range is best measured using a grid of catching or recording sites. Ideally, at least fifteen records are required from one animal (STICKEL, 1954) as below this number, the size of the range is not fully disclosed. How is home range measured from the accumulated location records? Several methods have been proposed that take varying account of the area peripheral to the recording points. The *minimum area* (Fig. 2–4a) is obtained by joining the outermost points of capture and

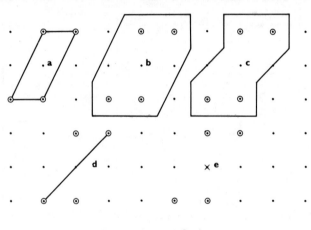

· Trap position   ⊙ Catch

Fig 2–4   Home ranges, (a) minimum area, (b) inclusive boundary strip, (c) exclusive boundary strip. Range length (d) and centre of activity (e)

calculating the area enclosed. The *boundary strip* is similar to minimum area but includes the addition of a strip half-way to the next trap; the *exclusive boundary strip* (Fig. 2–4c) has the nearest corners of the adjacent squares joined. *Range length* (Fig. 2–4d) is the distance between the most widely separated records and whilst not strictly a measure of the ground covered by the animal it can be a useful measure for comparative purposes. The same is true of the *centre of activity* (Fig. 2–4e); this is obtained by treating the recording sites as co-ordinates on a sheet of graph paper and calculating the vertical and horizontal means.

The foregoing methods are only suitable for animals inhabiting the ground surface. The home range of exclusively subterranean species such as moles and mole rats is defined by their underground network of tunnels and chambers. Their ranges can be obtained either by tracking through radioisotope tagging (GODFREY, 1957) or by the more laborious process of digging out the whole system (JARVIS and SALE, 1971).

2.5   Estimation of population density: live-trapping methods

How abundant are small mammals? Do their numbers change throughout the year? To answer these questions we can adopt various methods of population estimation. The method selected depends on various factors such as whether the population is to be surveyed regularly or whether it is to be a once-and-for-all count. Mammals are relatively big, compared for example to terrestrial invertebrates, and so large samples take considerable time to acquire. Also, their behaviour to traps can be very erratic. The 'trap shy' animal that avoids traps and the 'trap addict' which will enter them at every possible opportunity tend to bias the results.

When obtaining a figure for population numbers it is desirable that some statistical estimate of its accuracy accompanies it. The following examples fall in the two categories of providing (stochastic models) and not providing (deterministic models) this information. Several of these methods rely on the dispersion of traps in a grid followed by mark and recapture of the animals. The distance apart of the traps and the number at each trapping point will depend on the density and size of home range of the resident animals. The traps should be sufficiently numerous to ensure that no animal is ever excluded due to prior occupancy.

The simplest estimate of a population is derived from the Lincoln Index (LINCOLN, 1930). This can be expressed in the form:

$$N = x_1 x_2 / y$$

where N is the total population within the area, x_1 the number caught marked and released at the first sampling, x_2 the total number caught at the second sampling and y the number of the x_2 animals that had been marked at the first sampling. The method makes three assumptions: (i) that there is no population change through immigration, emigration, natality and mortality between the sampling periods (ii) all animals within the population are equally catchable and (iii) the marked animals mix freely with the rest of the population and are subsequently unaffected by the marking being neither easier nor more difficult to obtain than the unmarked ones. Even if the time interval between the two samplings is kept to a minimum it is difficult to see how these conditions can be met. The Lincoln Index has nevertheless provided a basis for development of more refined techniques involving the derivation of a variety of mathematical methods from theoretical considerations of successive marking and recaptures.

A computationally simple estimate based upon the increase in the proportion of marked animals in the population in successive catches has been proposed by HAYNE (1949). Three estimates are required for each day: (i) total number of captures (w), (ii) proportion of catch previously handled (y) and (iii) total number previously handled (x). The population is estimated as

$$\Sigma wx^2 / \Sigma wxy$$

A more elaborate deterministic model by LESLIE and CHITTY (1951) and LESLIE (1952) not only provided estimates of population density but also 'dilution rate' (the rate at which young trappable animals and immigrants were being added to the population) and 'death rate' (rate of disappearance due to death and emigration). But it was the application of the model to a 21-month field study of vole populations in mid-Wales (LESLIE, CHITTY and CHITTY, 1953) that highlighted the advantages and disadvantages of this technique. Meaningful parameters were obtained for the bank vole which responded well to the method. But it was with the field vole that the limitations of the method were exposed. For example, a zero dilution rate could be expected in the winter months when there was neither breeding nor immigration; in fact, the analysis revealed quite sizeable dilution values at this time. The authors attributed this to the different trappability of marked and unmarked animals in the non-breeding season, and furthermore recognized that this source of error could also have been operating, perhaps less obviously, during the breeding season. Similar problems are also inherent in HAYNE'S (1949) method. A more refined technique was developed by JOLLY (1965). He constructed a stochastic model which not only provided estimates of population, survival and dilution, but also the variances of these estimates. Elegant as this technique is, it does not overcome the problem of heterogeneity of trappability (TANTON, 1969).

Estimation of the untrappable segment of the population has been attempted by TANTON (1965). This he did by recording the number of times each animal was caught during a survey so that his data were finally summarized in the form a animals caught once, b twice, c three times, etc. These data were then fitted to a theoretical truncated negative binomial distribution (Fig. 2–5) from which it is possible, by extrapolation, to determine the number of animals caught 'no times'. One of the practical difficulties with this method is obtaining a large enough number of captures and recaptures in a short period of time but it obviously has very considerable attractions.

A simpler method used in the trapping of *Microtus* in North America has been proposed by HAYNE (in GOLLEY, 1960). This consisted of two

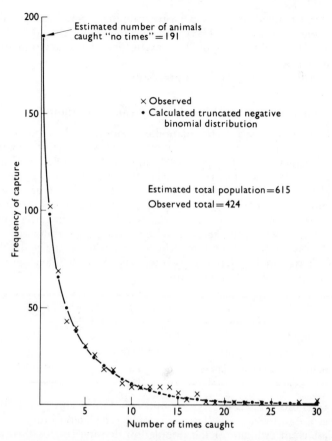

Fig. 2–5 Theoretical truncated negative binomial distribution and observed catches in the bank vole. (From TANTON, 1965)

100 m long trap lines at right angles to each other and crossing at their mid-points. The traps were spaced 2 m apart on each line. The lines were trapped alternate days for a total of six nights. The population density was estimated as bc/ad where a is the number of animals caught in common to both lines, b the average number of animals captured in each of the two lines (a and b exclude all animals dying in the traps in the trapping period), c average for the two lines of all captures including deaths due to trapping and d the effective area trapped.

The methods of LESLIE, JOLLY and TANTON are very much more sophisticated, and with further development and refinement can be expected ultimately to provide precise estimates of density. On the other hand they frequently demand extensive trapping programmes and time-consuming data analysis. There can be little doubt that the simpler methods such as those proposed by HAYNE give a less exact estimate but may have useful application, where the investigator's experimental situation does not impose rigid precision.

2.6   Estimation of population density: removal methods

A different approach to population estimation involves a permanent removal of animals from the area of survey. An effective and easily applicable method is the Standard Minimum Method of GRODZINSKI, PUCEK and RYSZKOWSKI (1966) which they applied to *Apodemus* in Poland. In this case snap traps were dispersed in a 16 × 16 grid with an interval of 15 m between each trapping point of two traps. An area of 5.76 ha was trapped with 512 traps. Cards with bait on them were placed at each trapping point for seven days before trapping commenced. They were then replaced with set traps which were visited at dawn and dusk; the catch removed and the traps reset. The data obtained were then summarized in the following form:

Day	1	2	3	4	5
No. caught	68	37	13	9	5
Total caught up to and including previous day	0	68	105	118	127

A regression line, its 95% confidence limits and its point of intersection of the abscissa (Fig. 2–6) can be calculated from these data. The last is an estimate of the population density. In this theoretical example the population is 135 and the formula of the line $Y = 68.81 - 0.51X$. Like TANTON'S (1965) method, this also estimates the uncaught portion of the population. For the success of this technique it is essential that the catch on the first day should be high and catches should steadily drop thereafter. This can be best achieved by ensuring that the animals are attracted through an adequate pre-baiting period and there are more than sufficient traps available. More recently PELIKAN (1971) has shown that the grid can provide an equally valid estimate if the number of trapping points is reduced to 8 × 8. In shrews a multicatch pitfall trap is more effective than snap traps in the summer months (PUCEK, 1969). The relatively short time taken to remove the animals minimizes the effects of immigration and emigration although there can be a small movement in from the edges of the grid towards the end of the trapping period.

In all the live and removal methods considered so far allowance must be made for the 'edge effect' of the grid. At its perimeter animals will be caught whose range may extend beyond the outermost traps or the periphery of the quadrats in which the traps are centred. This means that the trapping may be from an area greater (or possibly, but less likely smaller) than that being apparently covered. Adjustments therefore have to be made which take account of the home ranges and the rate of movement into the depopulated area. The larger the area under survey, providing it does not depart appreciably from a square shape, the less important the edge effect becomes.

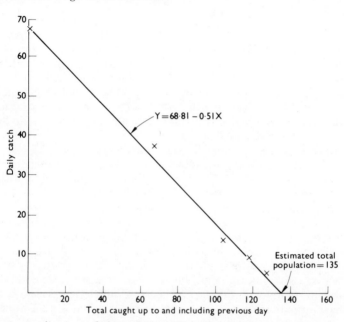

Fig. 2–6 Application of the Standard Minimum Method to estimate total population.

A completely different method of removal has been used by DIETERLEN (1967a) in eastern Zaire. He enclosed areas of between 500 and 1450 m² with a wall of corrugated iron sheeting a little under 1 m high. This was followed by removal of the vegetation and digging over the ground with the small mammals removed by hand as the work progressed. Some extremely large catches resulted. Extensive snap trapping also took place in similar areas to those where removal was in progress. Comparisons of the two sets of results indicated that the percentage numbers of each species caught were very dissimilar presumably as a result of the difference in the techniques and in each case one, but not the same, species was not caught at all. BELLIER (1967) working in Ivory Coast suggested a refinement by snap trapping within the enclosed area for four nights prior to the intensive hand removal. On theoretical grounds this method should be very efficient, but as BELLIER (1967) pointed out a large enough area must be selected to obtain representative coverage of animals. Its application is probably only worth while in situations where populations are high as the labour involved is very considerable and its use not warranted if the return is small.

2.7 Indices of abundance

Sometimes it is sufficient to have *relative* estimates of numbers in order to compare the changes in a particular place at the same time in different years. Also estimates of relative abundance in different habitats may provide the required information. In cases such as these an index of abundance is adequate. This can be obtained by using live traps in lines. SOUTHERN (1964) suggested the use of a trap line of 5 to 10 trapping points, 5 to 10 m apart with 5 traps at each point. If more than 80% of the traps are occupied their numbers should be increased. The line should be within a single habitat and the same type of trap used throughout. Pre-baiting is probably unnecessary and a single 24-hour trapping session minimizes the effects of trap addiction and disturbance of the population. A delay of three to four weeks is recommended between trappings. The number of traps set and the pattern of their dispersion can be changed in response to local conditions. Indices have provided useful information on the vertical distribution of *Apodemus* on the Island of Hirta, St. Kilda (BOYD, 1959) and in comparing ecodistributions on the Hebridean Islands and Scottish mainland (DELANY, 1961). But care must be exercised in their use. It is difficult to make meaningful comparisons between different species as their behaviour patterns may not be similar as well as between individuals of the same species at different times of the year.

2.8 Multiple use of methods

The use of more than one field method on a single population has obvious attractions and advantages. The second analysis can confirm at least some of the findings from the first and may also provide complementary information. It is also possible that the initial use of two methods may provide sufficient information to be able to disband one of the methods after a short time. HANSSON (1967) has demonstrated how this can be done through the combination of index trapping and population assessment. In this case a live-trap line was laid down containing five traps at each point and a 25 m interval between the 10 to 40 point centres comprising the line. Trapping and release took place with the total catch recorded. Immediately following this part of the exercise the traps were placed in two new trap lines parallel to and 22 m each side of the original index line. Average home range radius had previously been estimated at 25 m so that the two new lines enclosed an area completely embracing the home ranges of all the animals caught in the index line. The population was then assessed according to the Standard Minimum Method. Now the index figure can be related to density. If this exercise is repeated at the same time of year in several localities supporting different rodent densities a general relationship between index and abundance can be obtained.

An alternative use of assessment lines in conjunction with grid trapping has been suggested by SMITH et al. (1971). Immediately following intensive removal from the standard 5.76 ha grid, eight trap lines were laid, two parallel to each of the four axes of the grid. Each extended 60 m into the site of the grid and 120 m beyond its boundaries. These lines passed through three zones: (i) the innermost where there had been intensive trapping and removal: (ii) the intermediate, just beyond the perimeter of the original grid from which there had been partial depletion of the population; and (iii) the outermost region which was beyond the influence of the original trapping. From the twenty-two day catches obtained along the length of the assessment lines, paying particular attention to catching sites, it was possible to obtain information on the size of the area round the grid from which animals were attracted to the traps as well as the rate of movement into the area from which animals had been removed. In order to obtain the most suitable combination of methods MARTEN (1970) stressed the importance of selecting them so that the shortcomings of one is counterbalanced by the other. In censusing a population of *Peromyscus* in California he combined trapping and marking with tracking, using smoked-paper discs, and in so doing compensated for the shortcomings of trapping alone. The rationale behind this technique is complicated and the calculations elaborate but estimates were obtained of the numbers of animals within the census area together with the standard errors of the estimates.

Throughout this chapter the reader has been constantly reminded that there is a need to improve the methods used in field studies. Furthermore, the importance of obtaining large samples has been stressed as these permit rigorous statistical treatment. While recognizing the need for and usefulness of complex methodologies it is appropriate to draw attention to the tremendous gaps that exist in our knowledge of small mammals (see Appendix). Many of these can be filled with the help of a few traps and without recourse to complex numerical analyses. The opportunity exists then for all persons interested in natural history to make their contribution.

3 Life History Phenomena and Demography

If asked to describe the life history of a rodent we would probably include in our records its developmental changes from conception to death, the times when it reproduced and the number of young born on each occasion. In doing this not only is our knowledge of this animal increased but we also gain information on the way in which it is contributing to the next generation. The more animals we observe in this way the better able we are to generalize about the reproductive capabilities of the species. Ecologically, this is very important information as it provides an estimate of natality which is the population increase factor. Similarly, if we analysed our records on the ages at which animals died we could then work out life expectations. In this case we have data on mortality or the population decrease factor. Were this whole exercise repeated on the same species in a different locality we may well find that different conclusions would be reached about its natality and mortality. In other words these factors can be influenced by environmental conditions. Let us now consider some examples illustrating the range and tremendous adaptability of small mammal life histories and their relevance to population change.

3.1 Reproductive cycles and breeding seasons

One of the simplest reproductive cycles occurs in the European mole (*Talpa europea*) (GODFREY and CROWCROFT, 1960). Here, the female passes through oestrus from February to April, pregnancy from March to June and lactation from April to June (Fig. 3–1). There is some evidence of a very small proportion of females giving birth to a second litter, but there seems little doubt that the vast majority produce only one, i.e. are *monoestrous*. The male cycle fits this pattern very closely with the testes increasing to very large sizes in February and March at the time of mating, regressing appreciably in April and May (when mating producing a second litter may take place), and then still further into July. In June and July young animals become incorporated into the population and do not reproduce until the following year. The breeding season of the common shrew (*Sorex araneus*) in the British Isles extends from about April to October (BRAMBELL, 1935; CROWCROFT, 1957; SHILLITO, 1963). Like the mole, no reproduction has been recorded in the year of birth, but the adult females once they have started breeding continue producing litters throughout the summer, i.e. are *polyoestrous*.

The termination of reproduction results from the disappearance of these animals from the population and not the cessation of breeding followed by regression of the sexual organs. In the white-toothed shrew from the Isles of Scilly (*Crocidura suaveolens*) the reproductive pattern goes a step further in complexity. Not only is there breeding by the females born in the previous year; it can also occur in their year of birth. ROOD (1965) found in 1961 and 1962 that 35% and 13% respectively of

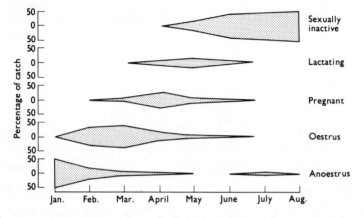

Fig. 3–1   The breeding season of the female mole *(Talpa europea)* in eastern England. (From GODFREY and CROWCROFT, 1960)

all the pregnant and/or lactating females were born in the same year as they bred. He also noted that breeding commenced much earlier in 1961 with pregnant females obtained in January. Breeding of shrews in their first year has also been recorded for *Blarina brevicauda* in the United States (PEARSON, 1945), *Sorex araneus* in continental Europe (PUCEK, 1960; STEIN, 1961) and *Crocidura hirta* in South Africa (MEESTER, 1963).

This last pattern is typical of most rodents, where often a sizeable proportion of the progeny breed within a few weeks of birth. The combination of early maturity, post-partum oestrus and a gestation period of 20 to 30 days affords dramatic population increase potential over a very short period. The birth of young generally coincides with an environmentally favourable time of year. In temperate climates this is from spring to autumn, in the seasonal tropics towards the end of the rains and in the aseasonal tropics all the year round (Fig. 3–2). These broad generalization do not hold for all species. An exception is seen in grassland in East Africa where several species of rodents inhabit a common area; of these the dark-bellied unstriped grass mouse (*Arvicanthis niloticus*) breeds continuously whilst the remaining species all display some degree of seasonal inactivity (DELANY and NEAL, 1969). The length of the breeding season can vary from year to year. ZEJDA (1962) and SMYTH (1966) have demonstrated how in the bank vole it may extend well into the winter in years of favourable food supply.

The flexibility in breeding by one species over a wide geographical area is well illustrated by the multimammate mouse (*Praomys natalensis*). In the Transvaal there is a long season extending through the southern hemisphere spring, summer and autumn (COETZEE, 1965). Approximately 3200 km north on the Uganda equator there are two short

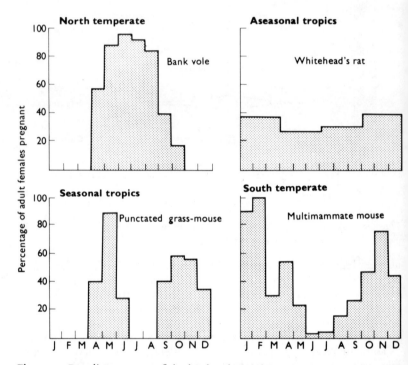

Fig. 3-2   Breeding seasons of the bank vole (*Clethrionomys glareolus*) in Britain, punctated grass-mouse (*Lemniscomys striatus*) in Uganda savanna, Whitehead's rat (*Rattus whiteheadi*) in tropical Malaysian rain forest and the multimammate mouse (*Praomys natalensis*) in South Africa. Data are only available on a quarterly basis for *R. whiteheadi*. (From BRAMBELL and ROWLANDS, 1936; NEAL, 1967; HARRISON, 1955 and COETZEE, 1965)

breeding seasons in May and June and October to November (NEAL, 1967) coinciding with the East African rains. Meanwhile, 5000 km away in Sierra Leone, West Africa, the main breeding is apparently from September to December (BRAMBELL and DAVIS, 1941). As temperate mountains are ascended the length of the breeding season tends to shorten. DUMIRE (1960) has demonstrated this with the white-footed mouse (*Peromyscus maniculatus*) in California. At up to 2300 m breeding occurred from spring to autumn, at 3200 m through spring and summer and at 4100 m in late summer and autumn.

3.2   Litter size

Mean litter size in rodents apparently ranges between the very low 1.1 and the remarkably high 12.1 (Table 3–1) although for most species it probably falls between 3 and 7. The number of young produced can vary according to the size and age of the mother; in the bank vole (BRAMBELL and ROWLANDS, 1936; KALELA, 1957) there is an increase with maternal size and in *Sorex* litter size declines towards the end of the

Table 3–1   Embryo numbers in gravid females.

Species	Mean no. embryos (range)	Locality	Authority
Shrews			
Crocidura suaveolens	3.0 (1–5)	Isles of Scilly, U.K.	ROOD, 1965
Sorex araneus	6.5 (1–10)	Wales, U.K.	BRAMBELL, 1935
Suncus murinus	1.8 (1–3)	Guam	DRYDEN, 1968
Rodents			
Otomys denti	1.1 (1–2)	Zaire and Rwanda	DIETERLEN, 1968
Praomys natalensis	12.1 (6–19)	Uganda	NEAL, 1967
Rattus exulans	4.3 (1–8)	Selangor	HARRISON, 1955
Clethrionomys rufocanus	6.0 (2–10)	Finland	KALELA, 1957

breeding season as a result of increased embryonic mortality (BRAMBELL, 1935; TARKOWSKI, 1957). Further intraspecific differences can be related to geographical and altitudinal factors. GODFREY and CROWCROFT (1960) reported mean litter sizes of 3.5 to 3.9 in the mole in Great Britain and 5 to 5.8 in Russia. In North America, LORD (1960) showed that by combining data on ten species of meadow mice (*Microtus*) there was for the genus as a whole a significant positive correlation between latitude and litter size. At 20°N the figure for *M. quasiater* was just over 2 and at 60°N in *M. pennsylvanicus* it approached 7. Litters of the white-footed mouse increase in size as the Colorado mountains are ascended with averages of 4.0 at 1675 to 1735 m and 5.4 at 2600 to 3400 m (SPENCER and STEINHOFF, 1968). Looking at these figures we may well ask ourselves how often do rodents with short breeding seasons have large litters? What are the mechanisms underlying such flexibility and adaptability? And what is their significance in relation to population regulation?

Embryonic mortality *in utero* may serve to produce a smaller number of young animals with improved chances of survival. This is borne out by BRAMBELL and ROWLANDS (1936) observations on the bank vole where there was between three and four times the loss in litters of 6 or more embryos than with those of 5 or less. A rather different, but nevertheless very interesting example of embryonic mortality has been observed by NEAL (1967) in the African unstriped grass mouse (*Arvicanthis niloticus*). Two populations a few miles apart had averages of 4.64 and 3.67 implanted embryos. However, higher uterine mortality in the former resulted in there being no significant difference in the number of live animals produced by each gravid female.

3.3 Postembryonic development

In most small mammals the young are born naked and blind. They then spend two to four weeks in a nest. During this time the eyes open, hair covers the body, responses to environmental stimuli develop, ear pinnae unfold, teeth erupt, toes separate, movement becomes more proficient and finally weaning occurs. At the time of leaving the nest they are replicas of the adult except for the immaturity of the gonads and the presence of juvenile pelage. A species with a relatively slow development is the tree mouse, *Dendromus melanotis* whose nests are constructed high up in the vegetation. In this small African mouse the incisors erupt after 10 days, the eyes do not open before 22 days and nest residence lasts until about 35 days (DIETERLEN, 1971). A nest-dwelling species like this is referred to as nidicolous and contrasts with the nidifugous condition where little or no time is spent in the nest. The spiny mice (*Acomys*) and laminate-toothed rats (*Otomys*) fall in the latter category (DIETERLEN, 1961, 1968). In *Otomys* the young are born with incisors erupted, eyes open, a good covering of hair and an ability to run actively. Weaning occurs at about 12 days and the young probably never reside in a permanent nest. An interesting intermediate development pattern is seen in the harsh-furred mouse (*Lophuromys flavopunctatus*). It is naked and blind at birth but by the fourth day the eyes open, there is a complete covering of dense hair and the young can walk about freely (DELANY, 1971). These changes are accompanied by extremely rapid growth (Fig. 3–3) with weight increases from 8 to 20 g in the first five days. *Lophuromys* and the two nidifugous genera produce large young and small litters.

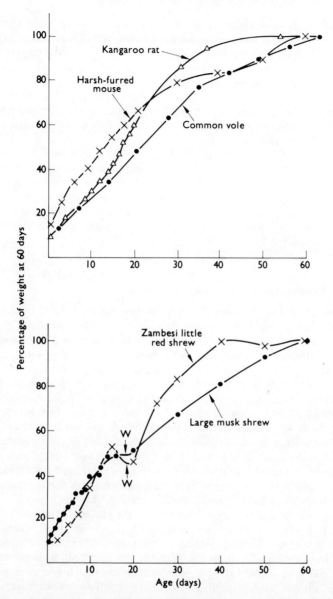

Fig. 3–3 Growth of rodents (Stephen's kangaroo rat, *Dipodomys stephensi*; common vole, *Microtus arvalis* and harsh-furred mouse, *Lophuromys flavopunctatus*) and insectivores (Zambesi little red shrew, *Crocidura hirta* and large musk shrew, *Suncus murinus*); w = time of weaning. (From LACKEY, 1967; REICHSTEIN, 1964; DELANY, 1971; MEESTER, 1963 and DRYDEN, 1968)

Growth is most rapid during the first 40 to 60 days. Thereafter such increase as occurs is much more gradual. In the shrews there may be a weight loss at weaning and during the winter (SHILLITO, 1963). In the rodents there can also be cessation of winter growth. This has been reported in *Clethrionomys* (ZEJDA, 1971) whilst actual recession is known to occur in the rat, *Sigmodon* (DAVIS and GOLLEY, 1963). But there can be great lack of uniformity in growth within a species. For example the time of year when they are born and the density of the population can influence the rate of growth.

The age of first reproduction is often influenced by environmental conditions. When unfavourable, breeding by what would otherwise be fecund animals, may be inhibited. The field vole, bank vole and wood mouse can all breed at 42 days (SOUTHERN, 1964) whereas other species such as the African forest rat (*Praomys morio*) and the Malaysian wood rat (*Rattus tiomanicus*) take 75 (EISENTRAUT, 1961) and 85 days (HARRISON, 1955) respectively to attain maturity.

3.4 Population structure and survivorship

The structure of a small mammal population is described by the number of animals in each age class such as weeks, categories of tooth wear (in which case we have an index rather than an absolute age) or simply the broad separation, which can be very useful, into juvenile, subadults and adults (Fig. 3–4). Considerable information can be derived from these data when they are supplemented by details of the life cycle. This has been demonstrated in laboratory colonies of the field vole (LESLIE and RANSON, 1940). Table 3–2 illustrates how two populations both of 1000 animals may have different growth characteristics. In the Malthusian distribution, where a large number of young animals are present, the birth rate is 0.1127 and the death rate 0.025. This means that within a unit of time of one week in a population of 1000 females 25 would die and 113 daughters would be born. In the life-table of a stationary population deaths would equal births.

Life expectancy or survival is a statistical estimate of the time to death and is likely to vary according to age and sex. In small rodents and insectivores life duration is usually short, seldom exceeding three years and for very many animals considerably less than one. A simple method of expressing survival is by the survivorship curve. This involves taking a large sample and recording how many animals are alive at various intervals of time. FLEMING (1971) working in tropical Panamanian forests found three species of rodent to have different shaped survivorship curves (Fig. 3–5). The rice rat (*Oryzomys capito*) had a high mortality in its early life and less so subsequently, the spiny pocket mouse (*Liomys adspersus*) disappeared at a steady rate and the longer lived

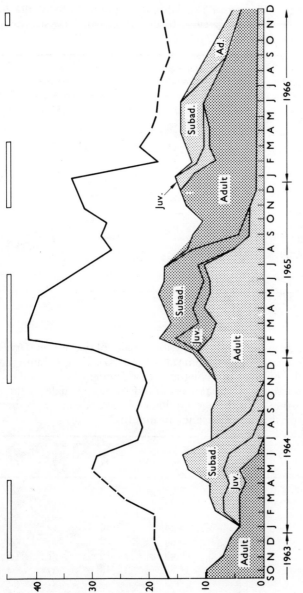

Fig. 3–4 Year classes (with uniform shading) and numbers of juveniles, subadults and adults of female *Rattus fuscipes* from 0.81 ha of rain forest in Queensland, Australia. Heavy line, total population; horizontal bars, breeding season. (From WOOD, 1971)

Table 3-2   Age structure of two laboratory populations of the field vole (*Microtus agrestis*). (From LESLIE and RANSON, 1940; courtesy *J. Anim. Ecol.*).

Age (weeks)	Stable or Malthusian	Life table or stationary
0–	577	235
8–	255	212
16–	107	178
24–	41	138
32–	14	97
40–	5	63
48–	1	38
56–	—	21
64–	—	10
72–	—	5
80–	—	2
88–	—	1
Total	1000	1000

Table 3-3   Life and fecundity table for females of the rice rat (*Oryzomys capito*) (modified from FLEMING, 1971). The l_x column shows the number surviving at the beginning of each month, m_x numbers of females produced by each female in the age group, q_x is the age specific mortality rate and e_x is the age specific life expectancy in months.

Age (months)	l_x	m_x	$l_x m_x$	q_x	e_x
0	100	0.00	0	28	2.9
1	72	0.00	0	29	2.8
2	51	0.00	0	29	2.8
3	36	0.99	36	28	2.8
4	26	0.99	26	27	2.7
5	19	0.99	19	32	2.4
6	13	0.99	13	31	2.3
7	9	0.99	9	32	2.2
8	7	0.99	7	29	1.6
9	5	0.99	5	40	1.1
10	3	0.99	3		

$$R_0 = 118$$

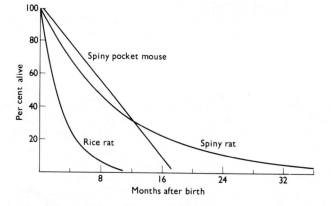

Fig. 3-5 Survivorship curves in the spiny pocket mouse (*Liomys adspersus*), rice rat (*Oryzomys capito*) and spiny rat (*Proechimys semispinosus*) in Panamanian forest. (From FLEMING, 1971; courtesy Museum of Zoology, University of Michigan)

spiny rat (*Proechimys semispinosus*) was intermediate in its survival during the first ten months.

Care must be exercised in deriving death rates at different ages (age specific mortality) from these curves. In the spiny rat, between the ages of 30 and 36 months the percentage of animals alive hardly changed so the number dying was very small and the mortality was then very low. Contrast this with the spiny pocket mouse; here about 6% of the *original* population died every month. Now, during the first month this would represent a mortality rate of 6%; but by the sixteenth month something like 94% of the original population would have died so that the mortality rate would then be approaching 100%. This sort of information is clearly summarized in the life table (Table 3-3). The inclusion of age specific fecundity, i.e. the mean number of females born to each age class, makes possible the calculation by summing the $l_x m_x$ column, of the numerical change in the population from one generation to the next. In this case a value of 118 indicates the replacement of 100 animals by 118 in the following generation providing conditions remain unaltered.

4 Populations

4.1 Densities and dynamics

The numbers of small mammals range from near zero towards the end of the adverse season to maxima, at times of cyclical peaks in voles, of 130–150/ha (Table 4–1). If the somewhat exceptional *Microtus* is discounted the maximum for temperate species appears to be nearer 40/ha. These figures in Table 4–1 do not necessarily represent the total small mammals in the habitats. The Californian desert shrub (CHEW and CHEW, 1970) also supported Merriam's Kangaroo rat (*Dipodomys merriami*) and the grasshopper mouse (*Onychomys torridus*) The former was about five times as numerous and the latter of similar density to the *Peromyscus*. There would probably be few other rodents in the situations studied by SOUTHERN, (1970), KREBS *et al.* (1973) and NEWSOME (1969). But in mixed grass and bush in Zaire, DIETERLEN (1967b) recorded twelve species of rodents which collectively had an *average* density in his eight samples, of the staggeringly high figure of 361/ha. Admittedly, these surveys were made in close proximity to cultivated fields into which the animals may have made feeding forays. Even so, these figures (which if anything underestimate rather than overestimate the numbers present) must be amongst the highest small mammal densities in the world.

In his extremely useful and unique twenty-year survey (Fig. 4–1) of

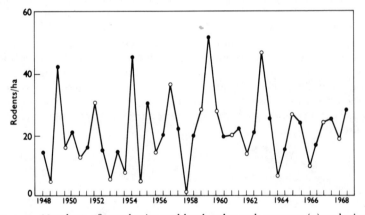

Fig. 4–1 Numbers of wood mice and bank voles each summer (o) and winter (●) in Wytham Wood, near Oxford from 1948 to 1968. (From SOUTHERN, 1970; courtesy Zoological Society of London.)

Species	Sampling routine	Density (with dates) nos/ha		Habitat	Authority
		Min.	Max.		
Sorex araneus	monthly, Nov. 65–Aug. 66	1.7 (Apr.)	6.9 (Nov.)	Deciduous woodland, England	BUCKNER, 1969
Sorex cinereus	monthly, May–Oct. 1952–5	2.0 (May–Jun.)	23.0 (Aug.–Oct.)	Coniferous bog, Manitoba	BUCKNER, 1966
Peromyscus eremicus	10 samples over 1 yr.	0.43 (Oct.)	3.3 (Jan.)	Desert, shrub, California	CHEW and CHEW, 1970
Apodemus sylvaticus	Apr., May or June and Dec., 1948–68	0.2 (Jun.)	19.6 (Dec.)	Deciduous woodland, England	SOUTHERN, 1970
Clethrionomys glareolus	Apr., May or June and Dec., 1948–68	0.8 (Jun.)	39.6 (Dec.)	Deciduous woodland, England	SOUTHERN, 1970
Microtus oeconomus	6 censuses over 2 years between May and Oct.	23 (Oct.)	131 (Jul.)	Mixed woodland, Poland	BUCHALCZYK and PUCEK, 1968
Microtus pennsylvanicus	about twice monthly, 1965–7	1 (Jan.)	150 (May)	Grassland, Indiana	KREBS et al., 1973
Mus musculus	almost monthly, 1963–7	0 (winter)	40 (Mar.)	Wheatfield, Australia	NEWSOME, 1969
Oenomys hypoxanthus	8 samples between Sept. 1963 and Mar. 1965		98.7 avg.	Grass and bush, Zaire	DIETERLEN, 1967b

Table 4–2 The productivity available to and utilized by small mammals. (From CHEW and CHEW, 1970; courtesy Ecological Society of America)

Locality and rodents	Av. biomass of mammals g/ha	Net above-ground primary productivity 10^8 J ha^{-1} yr^{-1}	Productivity available to mammals %	% of net consumed	% of available consumed	Authority
Larrea desert shrub						
Dipodomys (only)	590	238	4.2	1.5	3.5	CHEW and CHEW (1970)
All rodents and lagomorphs	1130	238	4.2	1.7	5.5	CHEW and CHEW (1970)
Old field Peromyscus	120	334	6.2*	0.8	13	ODUM et al., 1962
Old field Microtus	1900	661	100	1.6	1.6	GOLLEY, 1960
Fagetum forest						
Apodemus/Clethrionomys	730	1841	4.4	0.3	5.8	DROZDZ, 1966; GRODZINSKI, 1961

* Low estimate as only seeds were counted as available.

wood mice and bank voles in Wytham Wood, near Oxford SOUTHERN (1970), obtained twice-yearly estimates of the numbers in April or May or June and December. These were the times of year when populations were at their minima and maxima respectively. The numbers varied widely between the low of 1958 and the high of 1959 but on no occasion did either species approach the very high peak numbers of *Microtus* (see Table 4–1). There was no regular pattern of amplitude with uncharacteristic highs in the spring and early summer of 1952, 1957, 1959, 1960, 1963 and 1965 correlating with winter breeding in both species. The total populations tended to fluctuate about a mean of 20/ha and return to this figure within six months of a major displacement from it. The more extreme fluctuations between 1954 and 1959 could be associated with a major ecological change in the wood in 1954. In that year myxomatosis virtually eliminated rabbits and this resulted in an enforced change in the feeding habits of its main predators, particularly the fox. This change could have had considerable impact on the rodent populations which then took several years to readjust and return to a less violent fluctuation. If this were the case the precise mechanisms still remain undefined. The great value of a study like this is the way in which it provides an insight into these long-term fluctuations.

4.2   Population cycles

The dramatic and regular population fluctuations in voles and lemmings were first highlighted by ELTON in 1924. These cycles usually have peaks at intervals of 3 to 4 years although intervals of 2 and 5 years may also occur. It is the magnitude of the changes between the lows and highs that have excited the ecologist as much as the regularity of their occurrence. These microtine cycles have been recorded from the tundras of North America and Eurasia to as far south as California and New Mexico. However, this apparent southern limit may be related to lack of study further south rather than latitudinal limitation. In recent years, much work has been undertaken in North America with the following account largely derived from the work of KREBS *et al.* (1973) on *Microtus* in southern Indiana. Their research led them to look into intrinsic, i.e. the effects of individuals on each other, rather than extrinsic factors, e.g. food, predators. This approach stemmed from the earlier work of CHITTY (1967) who hypothesized that changes in genetic composition and behaviour through the cycle exerted important influences on the changes that took place within the cycle.

A typical *Microtus* cycle (Fig. 4–2) showed a rapid increase during autumn of year 1 with a peak attained by midwinter, then a small drop in spring and a recovery in summer of year 2. A steady but not very sharp decline occurred over the winter and spring followed by a very considerable drop in the summer of year 3. Accompanying this cycle

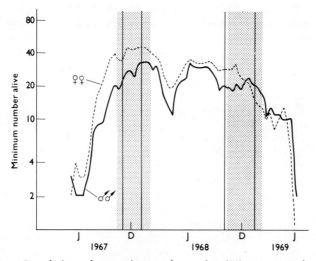

Fig. 4-2 Population changes in meadow mice (*Microtus pennsylvaticus*) in grassland in Indiana. Winter months are shaded; vertical lines separate summer breeding period from winter period. (From KREBS *et al.*, 1973; courtesy *Evolution*)

were several important changes within the population. There was a higher growth rate of individuals in the increasing population, a higher survival of adults and subadults during the peak phase accompanied by a dramatic increase in the death rate of juveniles, a lengthening of the breeding season in the increase phase with the attainment of sexual maturity at an earlier age and finally a shortening of the breeding season in the peak and decline phases. Higher survival and faster growth rates in the increasing population produced larger animals in the peak population.

Population changes were also compared in fenced and unfenced areas (Fig. 4-3). The fencing permitted the free movement of predators but no immigration or emigration by the voles. The maximum population in the enclosure was about treble that of the unenclosed area. This resulted in habitat destruction, overgrazing, symptoms of starvation and decline in the population. This was a considerable consequence of restricting dispersal. It was also found that from unenclosed areas most of the dispersal took place during the increase phase of the fluctuation and was least common in the decline. This means that most of the disappearance of animals during the peak and decline phases is within the habitat and not through movement outside it.

Attempts to relate these changes within the population to genetic phenomena have met with some success. There are definite indications that the genetic composition of dispersing individuals is not the same as the static ones so that the demographic events in *Microtus* are probably selective. But the way in which genetical selection operates remains unexplained. From behavioural studies it appears that males show significant changes in aggressive behaviour during the population cycle with individuals from peak populations being the most aggressive. Those dispersing during peaks were even more aggressive than those that remained. Increased aggression will thus tend to ensure optimal spacing of individuals.

From this extremely interesting research KREBS et al. (1973) have put forward a number of further ideas concerning these cycles and a general theory of their functioning (Fig. 4–4). The dispersers formed a group with a high reproductive potential as it contained a large proportion of young females in breeding condition and in this way would be able to exploit a new habitat rapidly. Those remaining behind made maximum use of the habitat without incurring the overuse that is evident when both components are forcibly kept together. These ideas on density-related changes in natural selection shed new light on mechanisms of population control and provide further useful, but in some ways still incomplete, information on the ways in which microtine rodents may regulate their numbers.

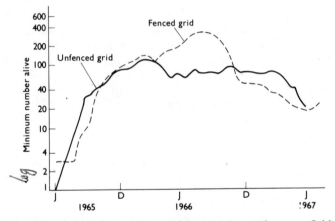

Fig. 4–3 Population changes in meadow mice in 0.8 ha grass fields. (From KREBS et al., 1973; courtesy Ecological Society of America)

4.3   Food supply

A few studies have been made in selected habitats on the quantity of food available to rodents in terms of its accessibility and palatability (Table 4–2). In general they make relatively small use (1.6 to 13%) of the total available to them over the year. A conclusion which might suggest appreciable underpopulation. However, closer examination of the data shed a different light on the situation. For example, CHEW and CHEW (1970) found that the mammals (including larger rodents and lagomorphs) of desert shrub collectively consumed 86.3% of the total available joules contained in seeds but only 2.3% in the browse. Furthermore, 75% of the seed joules were eaten by the kangaroo rat. As this animal is mainly a seed eater and these are produced seasonally, it is

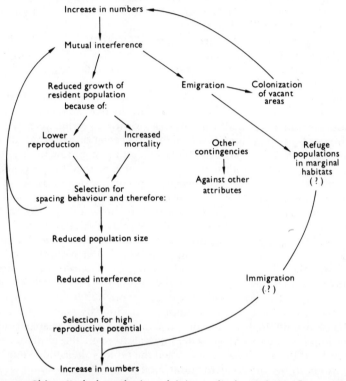

Fig. 4–4 Chitty–Krebs hypothesis explaining cyclical population fluctuations in small rodents. (From KREBS *et al.*, 1973; copyright 1973 by the American Association for the Advancement of Science)

quite probable that at certain times of the year it will be in danger of experiencing food shortages. Thus, although only 3.5% of the available food is eaten by the rodents the commonest species could still encounter times of inadequate supply.

The relationship between food supply and breeding season is both interesting and complex. SMYTH (1966) has shown how near Oxford winter breeding in *Clethrionomys* and *Apodemus* coincides with years of good acorn crops (so incidentally does that of *Microtus* which does not feed on acorns). From Kenya, TAYLOR (1968) has reported how exceptionally long rains in 1961 prevented farmers harvesting their crops and encouraged weed growth. These provided an unprecedented food source for four species of field rats which in turn attained pest proportions. NEWSOME (1970) added food to experimental field plots in Australia and thereby induced extremely high populations (up to 1100/ha) of the house mouse, *Mus musculus* within four months of initiating the experiments. In these examples an 'unexpected' additional food supply had a sudden and dramatic affect on the population, A more delayed influence has been described by CHEW and BUTTERWORTH (1964). Their study was made in the Mojave Desert, California from 1956 to 1958. Here, in this very unstable habitat extreme annual variations in rainfall and plant productivity are encountered. Usually, significant quantities of rain fall only in one or two months of the winter. In 1955, rainfall and hence plant productivity were high, in 1956 almost zero, in 1957 small with grasses growing a second year without fruiting and in 1958 rainfall was high and vegetation cover good. The numbers of *Dipodomys* were particularly high in 1956, low in 1957 and very low in 1958. In this example, it appears that the population response to weather and productivity was lagging by about a year.

Several species vary their diet during the year in response to changing availability. This is particularly obvious in the wood mouse (WATTS, 1968) which feeds mainly on seeds for most of the year and then switches to animal food, especially arthropods, during May and June (Fig. 4–5). All types of seeds are not uniformly consumed with sycamore and ash less attractive than acorn. After examining data on wood mouse numbers over a nineteen-year period WATTS (1969) concluded that food was only one of several regulatory factors. Poor survival of juveniles in summer was attributed to antagonistic behaviour by established adults (see p. 39) and higher autumn survival was due to changed adult behaviour. As there was considerable variation in the abundance of winter food the mice probably reached the limit of their food supply in one winter in two and only then was food presumed to be limiting. The work on *Microtus* has shown how in Indiana grassland even peak populations failed to reach the immediate food capacity of their habitats, and how other regulatory mechanisms operated before a situation of considerable habitat degradation occurred. It appears then that intrinsic mechanisms can be involved in adapting a population to a realistic equilibrium with available food. It would also seem reasonable to anticipate that carrying capacities in very different habitats would not be uniform for a particular species. This would then imply a relationship, so far undefined, between the intrinsic mechanism and the food supply.

The response of populations to differing food supplies has been examined by HOLLING (1959) in the feeding of the masked shrew (*Sorex*

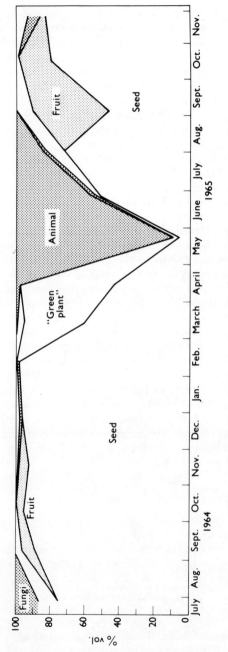

Fig. 4–5 Seasonal variation in food of the wood mouse in an English woodland. (From WATTS, 1968; courtesy *J. Anim. Ecol.*)

cinereus), the short-tailed shrew (*Blarina brevicauda*) and the deer mouse (*Peromyscus maniculatus*) on cocoons of the European pine saw fly (*Neodiprion sertifer*) in Canadian forests. The larvae fall to the ground in early June and immediately form a cocoon from which the adult emerges in September. A few emerge a year later. This additional food supply appears suddenly, can be of vast dimensions and most of it is only available for four months of the year. The number each animal consumes varies with the density of the cocoons and the species of mammal (Fig. 4–6a). This specific behavioural response to cocoon density is termed the functional response. HOLLING (1959) next estimated the numbers of small mammals in habitats with different cocoon densities (Fig. 4–6b) these being the numerical responses of the predators to prey density and finally he combined both responses to give an estimate of the extent of predation (Fig 4–6c). Various important points emerge. The numbers of small mammals do not show a linear relationship with increasing food. The percentage predation by each species shows an initial rise and subsequent decline with increased density. Thus up to some finite level of cocoon density, consumption by small mammals is directly density-dependent. Once again these animals can be seen to be taking advantage of increased food but are not fully exploiting the total available. In this case the food source is very temporary and other limiting factors may operate which bring the population into a satisfactory year round equilibrium.

The nature of the diet may be extremely important in initiating and terminating reproduction. Dietary protein levels will have to be maintained to ensure breeding and oestrogenic precursors must also be available. But so far little work has been attempted in this direction on wild rodents. NEGUS and PINTER (1966) stimulated oestrus in *Microtus* by feeding it on sprouted wheat and MILLAR (1967) found the oestrogenic activity of three African grasses varied in relation to rainfall. Even these preliminary observations very much implicate climate and food quality as regulatory factors.

4.4   Effects of fire and flood

These are two factors which could be expected when very severe to eliminate large numbers of animals from the areas where they occur, in a density-independent way. There have been few studies on the effects of fire and flood possibly because their occurrence is restricted to a limited

Fig. 4–6   Feeding responses of masked shrew (*Sorex cinereus*), short-tailed shrew (*Blarina brevicauda*) and deer mouse (*Peromyscus maniculatus*) to varying quantities of food supply in a Canadian woodland (**a**) Functional response, (**b**) Numerical response, (**c**) Combined functional and numerical response. (From HOLLING, 1959)

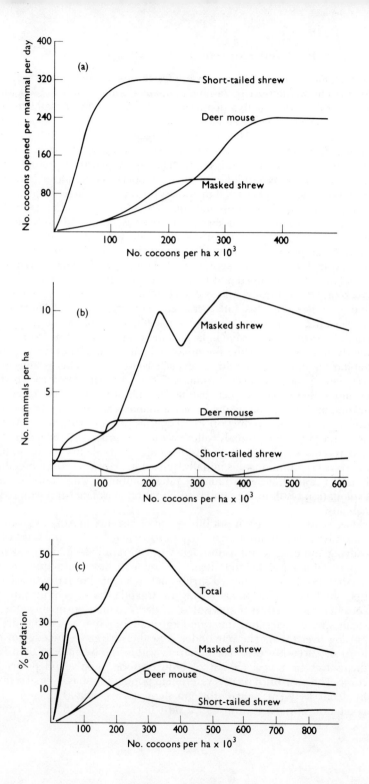

number of localities. The two examples chosen are both from tropical Africa. The Kafue river in Zambia annually floods a plain 235 km long and up to 40 km wide to a depth of 5 m. The heaviest rains fall between December and February and the flooded area reaches its greatest extent in about May or June. The floodplain is again uncovered well before the beginning of the next rains. The main breeding season of the small rodents is in the late rains (SHEPPE, 1972). As this precedes the inundation there is a build up in their numbers. As the water gradually penetrates the plain the small mammals form dense refugee populations round its margins and on islands or levées close to the main river. On one of these islands a rich vegetation cover was present in the late dry season and throughout the rains. Prior to the flood and during the early rains a high population of multimammate mice were present but by the time two-thirds of the area was shallowly flooded in March this population had dropped and those of the swamp rat (*Dasymys incomtus*), creek rat (*Pelomys fallax*) and two species of shrew (*Crocidura mariquensis* and *C. occidentalis*) had all increased very considerably. These four species all favour wetter situations than the multimammate mice and could be expected to reside in large numbers in swamp conditions. But the dynamic nature of the floodplain conditions does not provide the habitat stability that would be expected elsewhere. This results in a frequent change of species composition of a particular area. In this example there was further habitat modification through extensive grazing and trampling by the lechwe antelope between March and July causing a dramatic reduction in all species of small mammals. As the flood recedes the mammals follow it out and gradually recolonize the uncovered areas. But they do this in an unfavourable dry season so that the recovery in numbers is initially quite slow. However, once breeding is under way, high reproductive rates combined with mobility and exploitation result in the rapid recolonization of isolated and temporary habitats.

Fire is an irregular but not infrequent occurrence in African savanna. Burns are most common in the dry season when much of the vegetation is dying and easily combustible. They spread rapidly and are frequently superficial leaving behind them the unburnt bases and tufts of the grasses and the little-affected shrubs. BELLIER (1967) in Ivory Coast and DELANY (1964) and NEAL (1970) in Uganda have demonstrated a considerable post-burn survival of rodents. This is probably due to some species retreating into underground burrows when the fire is passing over them. The remainder are either driven to the edge of the burn or succumb to it. The survivors within the burnt area then find themselves in a new habitat of reduced food supply and protective cover. The post-burn recovery of rodent populations has been described in detail by NEAL (1970) for grassland in Uganda (Fig. 4–7). The vegetation also passes through a post-burn cycle which in itself must

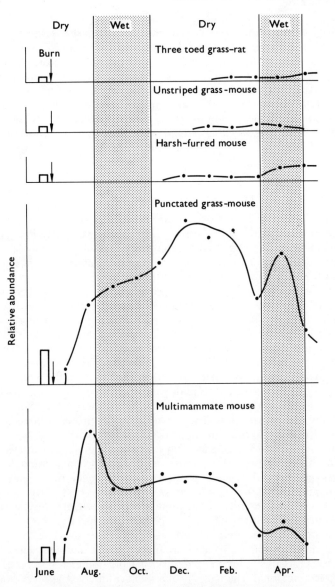

Fig. 4–7 The recolonization of savanna grassland in Uganda by rodents following a burn. (From NEAL, 1970): Three-toed grass rat (*Mylomys dybowskii*), Unstriped grass mouse (*Arvicanthis niloticus*), Harsh-furred mouse (*Lophuromys sikapusi*), Punctated grass mouse (*Lemniscomys striatus*), Multimammate mouse (*Praomys natalensis*).

influence the numbers and species of rodents present. Three rodent species did not reappear within the burnt area for at least five months. The punctated grass mouse and the multimammate mouse were present immediately after the burn and within a month their numbers increased to higher densities than immediately prior to the fire. In the former, maxima were obtained that were never recorded on comparable unburnt ground. These increases must have been largely due to immigration; the peak in numbers of the multimammate mouse is well in advance of the breeding season which normally accounts for the major additions to the population. These selective effects of burning are extremely interesting.

4.5 Intraspecific regulatory mechanisms

Intraspecific regulatory mechanisms have been known for some time in confined populations of house mice. In overcrowded conditions there is lower female fecundity, higher litter mortality due to cannibalism and high intra-uterine mortality with depressed lactation. But it was not until 1964 after ROWE, TAYLOR and CHUDLEY had examined populations in oat and wheat ricks that similar phenomena were established in the wild. At high densities the males had a lower survival rate than the females and after four months of the ricks being formed there was a steadily increasing movement of males (more than females) from the ricks. Embryonic resorption increased, as did aggressive behaviour amongst the males as indicated by a higher incidence of wounds. The main factors limiting populations were not the same in the two ricks. In oat ricks there was reduced fecundity in young breeding animals of both sexes as well as in the older females. In contrast there was a higher litter mortality in the wheat ricks. These differences can probably be related to qualitative dissimilarities in the two situations. In the oat there were quantities of weed offering nesting cover for aggregations of females who in this way protected themselves from aggressive males and thereby became non-fecund. The wheat offered no such cover, the females were mated and the young produced were not able to survive frequent disturbance and aggressive encounters. Here then are very sensitive regulatory mechanisms which change within a species in response to only a small habitat modification.

The investigation of rodents in more typical and less specialized situations reveals the occurrence of similar phenomena. SADLEIR (1965) examined seasonal changes in the behaviour of the deer mouse (*Peromyscus maniculatus*) in forests in British Columbia. He noted, as had been done for other species of rodents, that the numbers of young in the early summer population did not approach what would be expected from the number of embryos known to have been present in pregnant females a month or more previously. Laboratory and field experiments indicated that adults were very aggressive towards young animals in the breeding season and would tend to drive them out of their home areas at this time. The few that survived either displaced a low-aggression adult or remained inconspicuous, possibly through the occupancy of a very small home range. At the end of the breeding season the juveniles increased in numbers probably as a result of reduced adult aggression through declining testicular and ovarian activity. In the wood mouse BROWN (1966) has recognized the existence of a complex social organization in which a dominant male has an extensive home range which he carefully explores sector by sector. One or two females from each sector often accompany him but subordinate males inhabiting the area must give way to the dominant. There is then, a society with a hierarchical structure and a minimum of strife within it. However, fighting and other expressions of antagonistic behaviour can and do take place. They are most common towards newcomers to the area in the breeding season. The death or disappearance of the dominant results in his replacement by another or possibly more than one co-dominant.

These three species (the house mouse, the deer mouse and the wood mouse) display extremely important intraspecific regulatory mechanisms operating at two different phases of the population cycle. In *Mus* their main effect is experienced when populations are high whereas in the other two they operate to restrict increase in numbers when it would be expected that environmental conditions are propitious. On the other hand they do serve to disperse the population and encourage it to exploit new habitats at a favourable time of year.

4.6 Predation

PEARSON (1966) has made one of the few detailed quantitative studies of predation. This was on house mice (*Mus musculus*), meadow mice (*Microtus californicus*), harvest mice (*Reithrodontomys megalotis*) and gophers (*Thomomys bottae*) consumed by feral cats (*Felis domestica*), raccoons (*Procyon lotos*), grey foxes (*Urocyon cinereoargenteus*), striped and spotted skunks (*Mephitis mephitis* and *Spilogale putorius*) and possibly oppossum (*Didelphis marsupialis*) in an area of 14 ha of grasses and weeds in California. *Microtus* was the most abundant rodent and throughout the three years of study (1961–4) went through a complete population cycle of high maximum and low minimum.

As might be expected the predatory impact of the carnivores lagged behind the cycle of abundance of the *Micotus* (Fig. 4–8) with numbers of the meadow mouse available to each predator ranging from 72 in March 1962 to 540 in April 1963. This was due to an earlier and more rapid increase in the rodent than the carnivore population. The post-peak phase resulted in the carnivores chasing less and less prey so that the percentage of meadow mice being consumed by each predator was constantly increasing. At this time the predators also took relatively more harvest mice and gophers. As long as the predators could survive through this phase there was increased depletion of the rodents so that any recovery of their populations, particularly *Microtus* was not possible. Eventually the predators experienced food shortage, their numbers declined and meadow mice again increased to high peak populations. PEARSON (1966) considered that predators exercised control over the amplitude of the cycle through the maintenance of low densities over a long period. He did not believe they were responsible in any way for the post-peak decline. This is not in agreement with PITELKA *et al.* (1955) who consider that predatory birds and mammals were a primary factor causing the population crash of lemmings in Alaska. Their figures suggest a summer removal of at least 130 animals/ha.

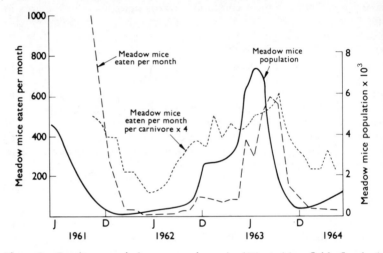

Fig. 4–8   Carnivore predation on meadow mice (*Microtus*) in a field of 14 ha in California. The number of *Microtus* eaten per month per carnivore is obtained by dividing the figures in the left hand column by four. (After PEARSON, 1966; courtesy *J. Anim. Ecol.*)

An interesting short-term study has been undertaken by SCHNELL (1968) on non-breeding cotton rats (*Sigmodon hispidus*) in Georgia, U.S.A. He introduced these animals into large enclosures of old field vegetation which differed from each other only in the numbers of rats introduced and their availability to predators in some enclosures and not others. He then measured survival of the animals in the different experiments. When not subjected to predation the rat populations showed low disappearance rates irrespective of whether there was a large or small number introduced initially. In contrast, under conditions of predation there was at the outset a steeper fall in the high density population than the low one (Fig. 4–9) but once the former had reached a certain level the reduction rate of its population declined. This has been attributed to it then attaining what has been called a 'predator-limited carrying capacity'. This density corresponded to that of the wild populations outside the study areas. SCHNELL (1968) suggests that at

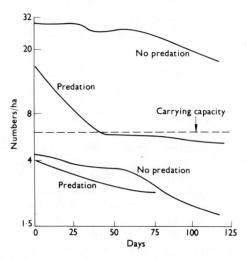

Fig. 4–9 Survival in non-breeding populations of the cotton rat (*Sigmodon hispidus*) with and without predation. Disappearance rates are slow when there is no predation and are greatest when there is predation and the initial densities are high. (After SCHNELL, 1968; courtesy The Wildlife Society)

densities below this level predation becomes a much less significant regulatory factor and therefore probably acts to a greater extent in the removal of 'surplus' animals from high populations. These experiments could have particular application to overwintering non-breeding populations and apparently suggest a different mode of predatory control than was found with *Microtus* in California.

5 Habitat Exploitation

5.1 Species diversity in different habitats

Small mammals, like any other group of animals, can collectively make maximum use of their habitat through a small number of species of catholic habits exploiting the wide range of resources available or alternatively by means of a larger number of species, several of more limited habits, specializing in the efficient exploitation of a particular resource. On the other hand, it may also be that the vegetational homogeneity and lack of floral diversity of an area would not be such as to encourage the evolution of a large number of species within it. Investigation of this phenomenon is a study in evolution, and whilst of very real interest and importance to the ecologist, the main concern here is to examine the results of this process in terms of the contemporary ecology of the various small mammal species.

The number of species of small rodents in various selected habitats (Table 5–1) ranges from 2 to 13. This is not large compared with some other groups of animals and probably fairly typically represents the sort of number to be expected in most natural situations. However, care must be taken not to generalize from this sort of data. The broomsedge-vine habitat is one of thirteen vegetational types recognized by GOLLEY *et*

Table 5–1 Numbers of species of small rodents in different habitats.

	Habitat	Locality	No. of small rodent species	Authority
Sub-arctic	Birch-crowberry lichen	Finnish Lapland	2	KALELA, 1957
Temperate	Broomsedge-vine	South Carolina	8	GOLLEY *et al.*, 1965
	Mixed woodland	northern England	3	ASHBY, 1967
Tropical	Moist forest	Panama	8	FLEMING, 1971
	Grassland savanna	Garamba, Zaire	12	VERHEYEN and VERSCHUREN, 1966
	Scrub-forest	Uganda	12	DELANY, 1971
	Alpine, over 3500 m	Mt. Kenya	5	COE and FOSTER, 1972

al. (1965) from a 78,000 ha. reserve in South Carolina. This supported the richest rodent fauna with the remaining twelve including a variety of forest, grassland and herbaceous habitats supporting from one to five species. Number of species does not of course give any indication of the number of animals present and it often happens that a small number of species may be numerous and the remainder ranging from common to infrequent. Animals in this last category may be typical of the habitat but have specialized requirements that prevent their appearance in large numbers or they may be characteristic of another type of habitat only occasionally entering the alien one. Examples in these two categories from English deciduous woodlands are the dormouse (*Muscardinus avellanarius*), with its preference for secondary growth of trees with an edible seed crop and the field vole (*Microtus agrestis*) which is found occasionally in grassy patches and herbage.

Detailed study is necessary to define the precise ecological requirements of each species. This embraces such aspects as the activity, food, microdistribution and behaviour to other species with the final picture emerging as a result of pooling this and other information. In most situations where this type of work has been undertaken in some detail, ecological differences emerge and it may well be that where no such separation between closely related species has been demonstrated that the depth of study has been inadequate.

5.2   Spatial distribution

The spatial distribution of an animal within a habitat, or its home range, can be described in terms of the situations it occupies whether these be biological such as trees, bushes and herbage or physiographic such as soils and rocks. These organic and inorganic components together provide a vertical stratification from the subsurface through the ground level to the high canopy of tropical forest so that an animal's distribution can be assigned to the layers it occupies as well as the type of situation. This definition can be taken a step further by indicating with which species of plants it may be found. In short, it is more precise to be able to say that our imaginary mouse inhabits the upper layers of oak and no other species of tree than to be able to say it occurs in the upper tree layer or just that it is arboreal. Unfortunately, detailed data are not available for a large number of species for which it is only possible, in most instances, to give a fairly general description of microdistribution.

The situation is further complicated by the animals not uniformly restricting themselves to a clear-cut and a regular distribution. Taking a relatively simple example, most workers agree that the two ubiquitous species of woodland small rodents in Britain, *Apodemus sylvaticus* and *Clethrionomys glareolus*, apparently have rather different habitat prefer-

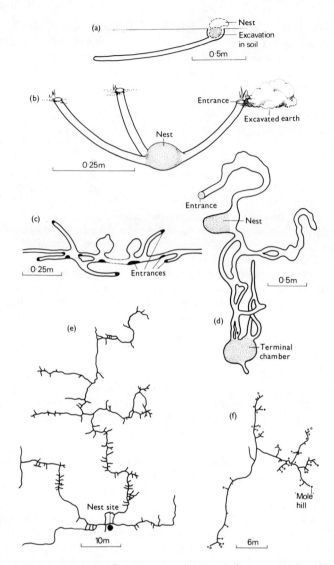

Fig. 5–1   Burrow systems of varying complexity. **(a)** Shaggy swamp rat (*Dasymys incomtus*), **(b)** Wood mouse (*Apodemus sylvaticus*), **(c)** Bank voles (*Clethrionomys glareolus*) within a few inches of the ground surface. **(d)** Bocage's gerbil (*Tatera valida*), **(e)** juvenile male European mole (*Talpa europea*) and **(f)** adult female African mole rat (*Heliophobius argenteus*); **(e)** and **(f)** are permanently subterranean. (After HANNEY, 1966; BARRETT-HAMILTON and HINTON, 1910–22; VERHEYEN and VERSCHUREN, 1966; GODFREY and CROWCROFT, 1960 and 'JARVIS and SALE, 1971; **a** and **f** courtesy Zoological Society of London)

ences. The former tends to be found more in open areas whereas the latter prefers situations with a good ground cover. But there are obviously many places within a wood where both species can be found. Furthermore, the low cover situation so readily favoured by the bank vole may vary very much from place to place in its floristic composition as well as its round the year permanence. This example only considers what occurs in a single dimension of the habitat.

Most individuals inhabit a well-defined area and have within it at least one nest where typically they rest for a large portion of the day, produce their young and possibly store food. It is the position of the nest that can often be responsible for extending the animal's range into another habitat dimension which it would not otherwise occupy. The ground dwelling *Apodemus* becomes partly subterranean as a result of the nest (with its access tunnels) being constructed underground. This pattern of two stratum occupation occurs in a large number of rodent species with their burrow systems ranging from the simple to the complex (Fig. 5–1). Other examples of different patterns of habitat utilization include the exclusively subterranean moles and mole-rats, the combined aquatic and terrestrial European water voles (*Arvicola* spp.), the arboreal and aerial pygmy scaly-tailed squirrels (*Idiurus* spp.) of Africa, the arboreal marmoset rat (*Hapalomys longicaudatus*) from Malaysia and the ground and field-layer inhabiting European harvest mouse (*Micromys minutus*).

Clearly, it is a combination of the behaviour, physical and physiological adaptations of a species that ultimately determine the distribution of its members. But it is only when the habitat is examined as a whole that the ecological differences between its component species can be seen most clearly. The richer the habitat is in species the more interesting the situation becomes. One such situation is the area of regenerating tropical forest studied by DELANY (1971) in central Uganda. His traps were set at various levels and were placed within trees, bushes, shrubs and herbaceous vegetation as well as on the ground. From the accumulated data it was possible to describe an interesting pattern of physical distribution (Fig. 5–2). Whilst this analysis tells us a good deal about the distribution of these animals there are obviously several overlaps. The next step would involve looking into other aspects of their ecology such as their feeding habits, distribution in relation to plant species and activity and thereby construct a comprehensive picture of the niche occupied by each species.

5.3   Activity

Activity can be studied in the laboratory as well as the field. In the former this can be achieved by direct observation or with the assistance of various pieces of apparatus which in principle depend on the movement of the animals activating a recording instrument. Activity

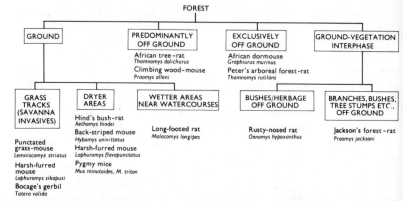

Fig. 5–2   The distribution of small rodents in a tropical forest in Uganda.
(After DELANY, 1971)

studies can often also be accompanied by observations on times of
feeding and drinking. Very useful results can be obtained in the field by
the regular and frequent visitation of live traps preferably with the
release of the catch on each occasion.

BROWN (1956) visited traps every two hours in a Berkshire woodland
over twenty-four-hour periods at different times of the year. Her results
(Fig. 5–3) show how *Apodemus* is essentially nocturnal with the midwinter pre-dawn and post-dusk peaks of activity not so obvious in the
short midsummer night whilst the diurnal *Clethrionomys* has a very
considerable mid-afternoon peak of activity in midsummer. Both
species adapt their activity to day length and furthermore display very
little overlap. In a laboratory study of shrews CROWCROFT (1954)
obtained strikingly different activity patterns from those found in these
two rodents. The daily pattern of the common shrew (Fig. 5–4) consisted
of about ten active periods, although these do to some extent overlap in
the accumulated results, with major peaks between 2000 and 0400 hours
and lesser peaks between 0700 and 1100 hours. The pygmy shrew had a
similar basic pattern but differed in being relatively more active during
the day.

Many tropical grasslands, scrub and forest support large numbers of
species. Here, day length varies little if at all, throughout the year and it
is unfortunate that no studies of comparable thoroughness to the
foregoing have been undertaken. In the grass-bush habitat of eastern
Zaire, DIETERLEN (1967b) trapped animals at 0800 hours, i.e. about one
hour after sunrise and 1600 hours or about three hours before sunset.
He was thus able to separate those animals that were clearly diurnal
from those that were active during the rest of the day. His results have
been expressed as the percentages obtained in the period preceding trap

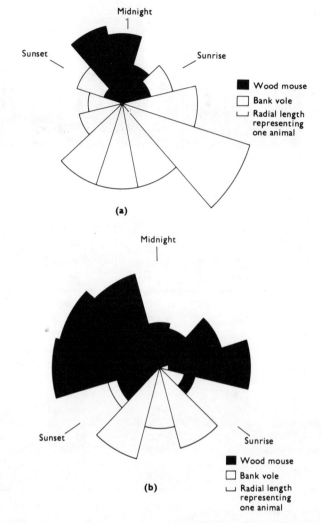

Fig. 5–3   The activity of the wood mouse and bank vole in southern England at midsummer (**a**) and midwinter (**b**). (From BROWN, 1956; courtesy Zoological Society of London)

examination (Table 5–2). These data undoubtedly mask small but important differences that may occur between species which apparently have virtually identical patterns. However, there still emerges a spectrum of different activity patterns affording an interesting comparison with the smaller fauna of the temperate woodland.

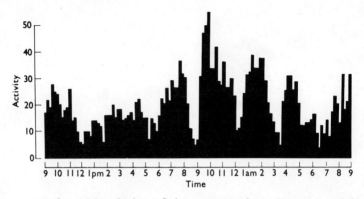

Fig. 5–4   Daily activity rhythm of the common shrew (*Sorex araneus*) in a laboratory illuminated from 0500 to 1700 hr. (From CROWCROFT, 1954; courtesy Zoological Society of London)

Table 5–2   Catches of rodents at different times in grass-bush habitat, eastern Zaire.

| Species | Total | Percentage of catch at | |
		0800 h	*1600 h*
Thamnomys dolichurus	15	100	—
Praomys natalensis	162	99.5	0.5
Praomys jacksoni	713	99	1
Dasymys incomtus	194	98	2
Mus minutoides	15	80	20
Otomys irroratus	87	79	21
Mus triton	184	70	30
Oenomys hypoxanthus	640	67	33
Lophuromys flavopunctatus	1099	39	61
Lemniscomys striatus	515	38	62
Pelomys fallax	62	35	65

5.4   Interspecific competition

When two rodents are found in different ecological situations do they maintain this separation passively or does each species actively prevent the other from exploiting a new sector of the habitat? The question is probably best answered by first considering to what extent one species will spread into a wider range of ecological conditions from those it normally occupies when given the opportunity. Continental islands often support faunas that are similar to but less rich in species than those of the nearby land mass. This situation is well illustrated in the

British Isles where there are frequently one or more of the commoner mainland species absent from the offshore islands. Island studies of the bank vole–field vole (*Microtus*) interaction suggest that in the absence of one of these species the other will invade the absentees habitat. FULLAGAR *et al.* (1963) found *Clethrionomys* widespread on Skomer and occupying typical *Microtus* habitat. On the Channel Islands of Jersey and Guernsey from which *Microtus* and *Clethrionomys* are respectively absent BISHOP and DELANY (1963) observed that both species lived in hedgebanks and meadows. The implication that one species is in some way instrumental in containing the other does not necessarily have uniform application. After extensive trapping in north-west Scotland, DELANY (1961) found the pygmy shrew no more abundant on islands from which the common shrew was absent than where it was present. The evidence is thus equivocal.

The *Microtus–Clethrionomys* problem has been studied experimentally in Canada. As in Britain, *Microtus* occurs in grassland and *Clethrionomys* in woodland. In one series of experiments a circular laboratory arena 2.45 m diameter had one half filled with grass (*Dactylis*) turfs and the other half with maple (*Acer*) saplings (GRANT, 1970). The voles were introduced in pairs of either the same or different species and their distributions and interactions observed. Behaviour was influenced by dominance relationships which tended to complicate the interpretation of the results. Even so it was possible to conclude that voles of different species had a mutually dispersive affect on each other as a result of aggressive interaction. If only one species was present it would occupy all the arena and not just the habitat type of its origin. In a larger scale field study (MORRIS and GRANT, 1972) in Saskatchewan an enclosure of 0.81 ha contained 0.59 ha of grassland and 0.22 ha of woodland. When *Clethrionomys* was removed from the latter *Microtus* entered in large numbers and when *Clethrionomys* was reintroduced the *Microtus* population declined. Mutual species exclusion was confirmed from a reciprocal experiment when *Microtus* was temporarily removed from grassland (GRANT, 1969). The mechanisms of exclusion and dispersion are not completely resolved by these experiments as MORRIS (1969) found both species co-existed in a Saskatchewan woodland under a protective layer of snow.

5.5 Feeding habits

The commonly held belief that the rodents are a group of exclusively herbivorous mammals is misconceived. Undoubtedly, very many species are herbivorous but one of the most adaptive features they display is their ability to feed on a variety of foods. This is manifested in two ways. Some species are very catholic in their tastes and eat a wide range of foodstuffs. Others are more specialized. Two species living in the British

Isles exemplify these types. Wood mice examined for 17 months from a Berkshire woodland ate the leaves, stems, flowers, fruits and seeds from at least twenty species of plants (some in very small quantities) as well as a diversity of animal material including caterpillars, centipedes, beetles, harvestmen and earthworms (WATTS, 1968). From the same general area, the field vole had a more restricted herbivorous diet obtaining its subsistence from seven species of grass (EVANS, 1973).

Looking at the picture more broadly LANDRY (1970) lists eighty-eight species of myomorph rodents that have been recorded as eating insects or flesh. Not included in his list, but among the more atypical diets, are the insectivorous white-bellied forest rat (*Colomys goslingi*) which enters streams for aquatic insects and crustacea and the predominantly ant- and termite-eating *Lophuromys sikapusi*, both from Africa (DELANY, 1974). The aquatic rodents of South America form another interesting and little studied group. Some are reputedly fish eaters although this has only been recorded in one genus (*Ichthyomys*) (THOMAS, 1893). Occasional flesh eating is not unusual. Records of birds, amphibia and reptiles being consumed are not uncommon. This tremendous diversity of feeding habits, with incidentally a remarkable uniformity of dentition ensures a very efficient collective utilization of food resources.

5.6 Energy dynamics

All organisms require energy to perform their body functions, reproduce and maintain themselves. In the case of rodents and insectivores this energy is derived from the plants and animals on which they feed. Instead of assessing the food in weight consumed we can use its energy content. This is measured in joules. The joule is very useful as it permits us to trace the use made of the food not only by the animal, but within the ecosystem as a whole. Supposing we look at the daily energy budget of an adult bank vole during the summer months (Fig. 5–5). Here we see that 88% of the energy consumed is assimilated and of this nearly 98% is used in keeping the body functioning and only the remaining 2% is put into the addition of new tissue.

What factors determine this pattern of energy flow? The immediate answer is very many. Ambient temperatures and the proportion of time the animal spends active modify metabolic rates and influence the respiratory heat loss. The *type* of activity makes different energy demands. DAVIS and GOLLEY (1963) have reported how the house mouse uses approximately 20% more energy during aggressive encounters than in investigative activity. The growing animal has greater tissue production while pregnant and lactating females of the bank vole require 58% more energy (KACZMARSKI, 1966) partly to maintain their increasing weight and partly to form tissue for the developing embryo. Energy requirements can vary between species as a result of differences in size and utilization rates. Heat production is correlated with body size in mature mammals with the metabolic rate of small ones being relatively greater than large ones. GEBCZYNSKI (1966) found that in winter the yellow-necked mouse (*Apodemus flavicollis*) utilized 1731 J/g body weight/day as compared to 2050 J/g/day in the bank vole (GORECKI, 1968). The mouse with an average weight of 29 g then required 50 200 J/day and the 21 g vole 42 700 J/day.

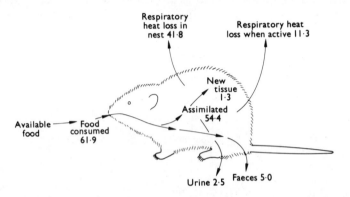

Fig. 5–5 A reconstructed daily energy budget of a bank vole weighing 23 g living in the laboratory during the summer. (From DROZDZ, 1968 and GORECKI, 1968); figures are × 10³ J. MEESE (1971) draws attention to a much higher energy intake of approximately 128 × 10³J under natural conditions

Supposing these energy exchanges are now looked at in the population rather than the individual. The energy contained within the animals at a particular time is the standing crop and the total content over twelve months is the annual production. To obtain this latter figure it is necessary to know how many animals are produced during the year and their average weight at death. Then the calorific values of their tissues can be calculated. For dried mammal tissue this is about 20 900 J/g although for very fat animals it can be as high as 39 700 J/g. A good example of this type of study was undertaken by GOLLEY (1960) on the vegetation–meadow mouse (*Microtus pennsylvanicus*)–weasel (*Mustela rixosa*) food chain in a field in Michigan (Fig. 5–6). Here the mice consumed less than 2% of the available food and the weasels about 31%. There was a considerable immigration of mice into the field which increased production to more than three and a half times what was produced locally.

This unidirectional food chain considers a situation where only two species of mammal were abundant. When more species are present the pattern of energy utilization is more complex as in the Californian desert shrub (CHEW and CHEW, 1970). A summary of the plant production consumed by the commonest species (*Dipodomys*) and all thirteen species of rodents and lagomorphs inhabiting the area is given in Table 4–2 (p. 27). Detailed examination of the density, biomass and energy budget (=assimilated energy) of the more important species is very revealing (Fig. 5–7). The kangaroo rat (*Dipodomys*) is clearly an important species. The jack rabbit (*Lepus californicus*) and the cottontail (*Sylvilagus auduboni*) occur in small numbers and utilize relatively less energy than the biomass suggests as a result of their large size. Eighty-three per cent of the total energy assimilated by these mammals is

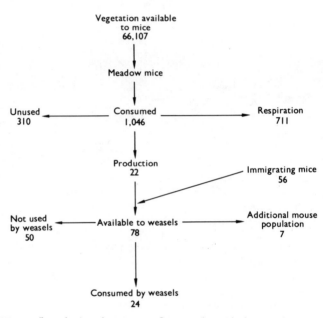

Fig. 5–6   Energy flow during the course of a year through the meadow mouse (*Microtus pennsylvanicus*) section of the food chain in a field in Michigan. The figures are × 10⁶J/ha/year. (From GOLLEY, 1960; courtesy Ecological Society of America)

accounted for by three species. Very interestingly one is a granivore (*Dipodomys*, 55%), one a browser (*Lepus*, 22%) and one an insectivore (*Onychomys*, 6.5%). The remaining species, individually, make relatively little demand on the environment and collectively only account for 16.5% of the assimilated energy.

Studies like this make considerable use of various demographic data on the populations including reproductive and mortality rates, densities and fluctuations as well as characteristics of the animal, including its growth rate and food consumption. The incorporation of so much information into a single study enables the formulation of the place rodents occupy in the ecosystem relative to each other. If their collective requirements were measured against those of other animal groups a more embracing picture of ecosystem dynamics would emerge.

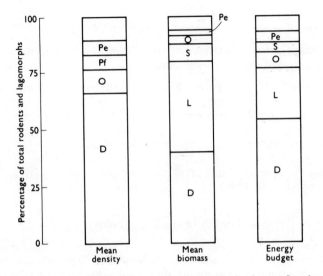

Fig. 5–7   Population densities, biomass and energy budgets of rodents and lagomorphs in the Californian desert shrub over a period of one year. Note how the jack rabbit and cottontail have very low population densities but by virtue of their large size comprise a very substantial proportion of the biomass and a slightly lower but nevertheless considerable proportion of the energy budget. D—Merriam's kangaroo rat (*Dipodomys merriami*), L—Jack rabbit (*Lepus californicus*), O—Grasshopper mouse (*Onychomys torridus*), Pe, Pf—Desert mice (*Peromyscus eremicus* and *P. flavus*), S—Cottontail rabbit (*Sylvilagus auduboni*). (From CHEW and CHEW, 1970; courtesy Ecological Society of America)

Appendix

Suppliers of traps

The traps previously described in the text can be purchased direct from the manufacturers as follows:

Longworth: The Longworth Scientific Instrument Company, Abingdon, Berkshire, England.

Sherman: Dr. H. B. Sherman, P.O. Box 683, Deland, Florida, 32720, U.S.A.

Havahart: Mr. F. B. Conner, P.O. Box 551, Ossining, New York, 10562, U.S.A.

Museum Special: Animal Trap Company of America, Lititz, Pennsylvania, U.S.A.

Guides to the identification of small mammals

The guides and keys listed below do not represent the total available. They are intended to introduce the reader with little previous experience of small mammals to the species that are present in many areas of the world as well as enabling or helping him to identify the animals he may collect. They are all in the English language.

In preparing this list it soon became evident that for some areas such as Europe and North America there is a reasonable quantity of bibliographic material. In contrast, the literature for many other regions, particularly the tropics, is very scant. Because of this reference has had to be made to a few works that are not readily obtainable as there is no comparable and available alternative.

Africa (East and Central)

ANSELL, W. F. H. (1960). *Mammals of Northern Rhodesia*, Govt. Printer, Lusaka.

DELANY, M. J. (1975). *The Rodents of Uganda*, Brit. Mus. (Nat. Hist.), London.

KINGDON, J. (1974). *East African Mammals* II. Academic Press, London.

SMITHERS, R. H. N. (1971). *The Mammals of Botswana,* Trustees of the National Museums of Rhodesia, Salisbury.

Africa (West)

ROSEVEAR, D. R. (1969). *The Rodents of West Africa*, Brit. Mus. (Nat. Hist.), London.

America (North)

Numerous mammal guides ranging from the introductory to the detailed and comprehensive are available for the United States and Canada. Many of these are regional in that they refer to only one state or a particular area. However, a useful general guide which is also comprehensive in its coverage is W. H. BURT (1952), *A Field Guide to the Mammals*, Houghton Mifflin, Boston.

Asia (East)

TATE, G. H. H. (1947). *Mammals of Eastern Asia*, Macmillan, New York.

Asia (Malaya)

MEDWAY, LORD (1969). *The Wild Mammals of Malaya*, Oxford U.P., Kuala Lumpur.

Australia

RIDE, W. D. L. (1970). *A Guide to the Mammals of Australia*, Oxford U.P., New York.

TROUGHTON, E. (1966). *Furred Animals of Australia*, 8th ed., Angus and Robertson, Sydney.

Europe

VAN DEN BRINCK, F. H. (1967). *A Field Guide to the Mammals of Britain and Europe*, Collins, London.

CORBET, G. B. (1964). *The Identification of British Mammals*, Brit. Mus. (Nat. Hist.), London.

Middle East

HARRISON, D. L. (1972). *The Mammals of Arabia III. Lagomorpha, Rodentia*, Ernest Benn, London.

Pacific World (including East Indies)

CARTER, T. D., HILL, J. E. and TATE, G. H. H. (1946). *Mammals of the Pacific World*, Macmillan, New York.

Fields in need of further study

The contributions that may be made by the local Natural History Society, the non-professional biologist, the undergraduate and the sixth former vary tremendously with the geographical location of where the work is to be undertaken. This is simply due to the fact that throughout the world as a whole there has been a great unevenness in the scientific study of small mammals. In North America and Europe both the professional and amateur biologist have made considerable strides in the advancement of knowledge on these animals. In contrast, in the developing world, where faunas are often richer, progress has been appreciably slower.

Dealing with the latter first, a great deal of information can be derived from the preparation of faunal lists from selected habitats particularly if these are accompanied by records of the distribution of individual species in relation to such features as vegetation, water and rock outcrops.

Little is known of the food and feeding habits of many of these animals and much can be revealed by offering caged animals a range of choice of their natural foods on a cafeteria basis. Information on activity can be obtained by observing captive animals or visiting traps regularly, e.g. six-hour intervals, over a 24-hour period. By breeding animals in captivity considerable light can be shed on their post-embryonic development, age to maturity and litter size.

In proposing topics within the British Isles consideration has to be given to what has already been accomplished and what is practicable with limited time and resources. With these considerations in mind the following lines of investigation are suggested:

1   Much useful information derives from a precise knowledge of the geographical distribution of animals. In order to provide an accurate record the Mammal Society of the British Isles initiated the accumulation of this information. Records have been collected on the basis of the 10 km squares of the National Grid and are now stored at the Biological Records Centre, Monk's Wood, Abbots Ripton, Huntingdon. Record cards and an explanatory leaflet can be obtained from this address. Distribution maps of British Mammals are given in CORBET (1971, *Mammal Review* **1** (4/5, 95).

2   Little is known of the ecology of small mammals in marginal habitats such as heather moors, sand dunes, mountains and upland grassland. Which species are to be found in these areas? Are they resident throughout the year? What do they feed on? At what time of year do they reproduce?

3   Few studies have been made on the year to year fluctuations in numbers of rodents. A yearly survey preferably in late autumn or early winter when populations are high, always conducted in the same way and at the same place could indicate the occurrence of cyclical fluctuations. This could be achieved by index trapping and would not necessarily involve determination of population densities. Traps would have to be set in sufficient numbers to adequately reflect the density of the animals.

4   To what extent do small rodents, particularly *Clethrionomys* and *Apodemus*, utilize situations above ground level? Little is known of the extent to which they exploit the tree, bush and shrub habitat. Trapping in these situations could be rewarding.

5   Numerous small islands within the British Isles have not had their small mammal faunas investigated. Many of those that have, have only been examined very superficially. What habitats do small

mammals exploit on small islands where the fauna is often reduced?

6   Information is very sparse on the nature of small mammal burrow and runway systems. What form do they take? How variable is their structure? How permanent are they? To what extent is there multiple occupancy?

7   There are indications that the onset and termination of breeding may vary from one geographical location to another. Regular trappings in one area, that could provide information on the time when first and last reproduction take place each year, over a period of several years, would be useful. The accumulated data from several widely separated places would help complete the general picture.

The Mammal Society of the British Isles provides a forum for the discussion of work in progress on all groups of mammals. It holds regular meetings and provides members with its two publications: *Mammal Review* and *Notes from the Mammal Society*. Further particulars can be obtained from the Assistant Secretary to the Mammal Society, c/o Harvest House, 62 London Road, Reading, Berkshire, RG1 5AS.

References

ASHBY, K. R. (1967). *J. Zool.*, **152**, 389.

BARRETT-HAMILTON, G. E. H. and HINTON, M. A. C. (1910–21). *A History of British Mammals*, Gurney and Jackson, London.

BELLIER, L. (1967). *Terre Vie*, **3**, 319.

BISHOP, I. R. and DELANY, M. J. (1963). *Mammalia*, **27**, 99.

BOYD, J. M. (1959). *Proc. zool. Soc. Lond.*, **133**, 47.

BRAMBELL, F. W. R. (1935). *Phil. Trans B*, **225**, 1.

BRAMBELL, F. W. R. and DAVIS, D. H. S. (1941). *Proc. zool. Soc. Lond.*, **111**, 1.

BRAMBELL, F. W. R. and ROWLANDS, I. W. (1936). *Phil. Trans B*, **226**, 71.

BROWN, L. E. (1956). *Proc. zool. Soc. Lond.*, **126**, 549.

BROWN, L. E. (1966). *Symp. zool. Soc. Lond.* **18**, 111.

BUCHALCZYK, T. and PUCEK, Z. (1968). *Acta theriol.*, **13**, 461.

BUCKNER, C. H. (1966). *J. Mammal.*, **47**, 181.

BUCKNER, C. H. (1969). *J. Mammal.*, **50**, 326.

BURT, W. H. (1943). *J. Mammal.*, **24**, 354.

CHEW, R. M. and BUTTERWORTH, B. B. (1964). *J. Mammal.*, **45**, 203.

CHEW, R. M. and CHEW, A. E. (1970). *Ecol. Monogr.*, **40**, 1.

CHITTY, D. (1967). *Proc. Ecol. Soc. Aust.*, **2**, 51.

CHITTY, D. and KEMPSON, D. A. (1949). *Ecology*, **30**, 536.

COE, M. J. and FOSTER, J. B. (1972). *J. East Africa Nat. Hist. Soc. Nat. Mus.*, **131**, 1.

COETZEE, C. G. (1965). *Zool. afr.*, **1**, 29.

CROWCROFT, P. (1954). *Proc. zool. Soc. Lond.*, **123**, 715.

CROWCROFT, P. (1956). *Proc. zool. Soc. Lond.*, **127**, 285.

CROWCROFT, P. (1957). *The Life of the Shrew*, Reinhardt, London.

DAVIS, D. E. and GOLLEY, F. B. (1963). *Principles in Mammalogy*, Reinhold, New York.

DELANY, M. J. (1961). *Proc. zool. Soc. Lond.*, **137**, 107.

DELANY, M. J. (1964). *Revue Zool. Bot. afr.*, **70**, 129.

DELANY, M. J. (1971). *J. Zool.*, **165**, 85.

DELANY, M. J. (1975). *The Rodents of Uganda*, British Museum (Nat. Hist.), London.

DELANY, M. J. and DAVIS, P. E. (1960). *Proc. zool. Soc. Lond.*, **136**, 439.

DELANY, M. J. and NEAL, B. R. (1969). *J. Reprod. Fert.* Suppl. **6**, 229.

DIETERLEN, F. (1961). *Z. Säugetierk.*, **26**, 1.

DIETERLEN, F. (1967a). *Acta trop.*, **24**, 244.

DIETERLEN, F. (1967b). *Zool. Jb. Syst.*, **94**, 369.

DIETERLEN, F. (1968). *Z. Säugetierk.*, **33**, 321.

DIETERLEN, F. (1971). *Säugetierk. Mitteil.*, **19**, 97.

DROZDZ, A. (1966). *Acta theriol.*, **11**, 363.

DROZDZ, A. (1968). *Acta theriol.*, **13**, 367.

DRYDEN, G. L. (1968). *J. Mammal*, **49**, 51.

DUMIRE, W. W. (1960). *Ecology*, **41**, 174.

EISENTRAUT, M. (1961). *Bonn. zool. Beitr.*, **12**, 1.

ELTON, C. S. (1924). *Brit. J. exp. Biol.*, **2**, 119.

EVANS, D. M. (1973). *J. Anim. Ecol.*, **42**, 1.

FLEMING, T. H. (1971). *Misc. Publ. Mus. Zool., Univ. Michigan*, **143**, 1.

FULLAGAR, P. J., JEWELL, P. A., LOCKLEY, R. M. and ROWLANDS, I. W. (1963). *Proc. zool. Soc. Lond.*, **140**, 295.

GEBCZYNSKI, M. (1966). *Acta. theriol.*, **11**, 391.

GENTRY, J. B., SMITH, M. H. and BEYERS, R. J. (1971). *Ann. Zool. Fennici*, **8**, 17.

GODFREY, G. K. (1955). *Ecology*, **36**, 678.

GODFREY, G. K. (1957). *Proc. zool. Soc. Lond.*, **128**, 287.

GODFREY, G. K. and CROWCROFT, P. (1960). *The Life of the Mole*, Museum Press, London.

GOLLEY, F. B. (1960). *Ecol. Monogr.*, **30**, 187.

GOLLEY, F. B., GENTRY, J. B., CALDWELL, L. D. and DAVENPORT, L. B. (1965). *J. Mammal.*, **46**, 1.

GORECKI, A. (1968). *Acta theriol.*, **13**, 341.

GRANT, P. R. (1969). *Can. J. Zool.*, **47**, 1059.

GRANT, P. R. (1970). *Anim. Behav.*, **18**, 411.

GRODZINSKI, W. (1961). *Bull. Acad. Polonaise Sci.* II **9**, 493.

GRODZINSKI, W., PUCEK, Z. and RYSZKOWSKI, L. (1966). *Acta. theriol.*, **11**, 297.

HANNEY, P. (1965). *J. Zool.*, **146**, 577.

HANSSON, L. (1967). *Oikos*, **18**, 261.

HARRISON, J. L. (1955). *Proc. zool. Soc. Lond.*, **125**, 445.

HAYNE, D. W. (1949). *J. Mammal.*, **30**, 399.

HOLLING, C. S. (1959). *Canad. Entomol.* **41**, 293.

JARVIS, J. U. M. and SALE, J. B. (1971). *J. Zool.*, **163**, 451.

JEWELL, P. A. (1966). *Symp. zool. Soc. Lond.*, **18**, 85.

JOLLY, G. M. (1965). *Biometrika*, **52**, 225.

KACZMARSKI, F. (1966). *Acta theriol.*, **11**, 409.

KALELA, O. (1957). *Ann. Acad. Sc. Fennicae* A. IV, **34**, 1.

KREBS, C. J., GAINES, M. S., KELLER, B. L., MYERS, J. H. and TAMARIN, R. H. (1973). *Science*, **179**, 35.

LACKEY, J. A. (1967). *J. Mammal.*, **48**, 624.

LANDRY, S. O. (1970). *Quart. Rev. Biol.*, **45**, 351.

LESLIE, P. H. (1952). *Biometrika*, **39**, 363.

LESLIE, P. H. and CHITTY, D. (1951). *Biometrika*, **38**, 269.

LESLIE, P. H., CHITTY, D. and CHITTY, H. (1953). *Biometrika*, **40**, 137.

LESLIE, P. H. and RANSON, R. M. (1940). *J. Anim. Ecol.*, **9**, 27.

LINCOLN, F. C. (1930). *U.S. Dept. Agriculture Circ.*, **118**, 1.

LINN, I. J. and SHILLITO, J. (1960). *Proc. zool. Soc. Lond.*, **134**, 489.

LORD, R. D. (1960). *Americ. Midl. Nat.*, **64**, 488.

LOWE, V. P. W. (1971). *J. Anim. Ecol.*, **40**, 49.

MARTEN, G. G. (1970). *Ecology*, **51**, 291.

MEESE, G. B. (1971). *J. Zool.*, **163**, 305.

MEESTER, J. (1963). *Transvaal Mus. Mem.*, **13**, 1.

MILLAR, R. P. (1967). *Rhod. Zamb. Mal. J. Agric. Res.*, **5**, 179.

MILLER, L. S. (1957). *Ecology*, **38**, 132.

MORRIS, R. D. (1969). *J. Mammal.*, **50**, 291.

MORRIS, R. D. and GRANT, P. R. (1972). *J. Anim. Ecol.*, **41**, 275.

NEAL, B. R. (1967). *The Ecology of Small Rodents in the Grassland Community of the Queen Elizabeth National Park, Uganda*, Ph.D. Thesis, University of Southampton.

NEAL, B. R. (1970). *Revue Zool. Bot. afr.*, **81**, 29.

NEGUS, N. C. and PINTER, A. J. (1966). *J. Mammal.*, **47**, 596.

NEWSOME, A. E. (1969). *J. Anim. Ecol.*, **38**, 341.

NEWSOME, A. E. (1970). *J. Anim. Ecol.*, **39**, 299.

ODUM, E. P., CONNELL, C. E. and DAVENPORT, L. B. (1962). *Ecology*, **43**, 88.

PEARSON, O. P. (1945). *Americ. Midl. Nat.*, **34**, 531.

PEARSON, O. P. (1966). *J. Anim. Ecol.*, **35**, 217.

PELIKAN, J. (1971). *Ann. Zool. Fennici*, **8**, 3.

PETRUSEWICZ, K. and ANDRZEJEWSKI, R. (1962). *Ecol. Pol.* A, **10**, 85.

PITELKA, F. A., TOMICH, P. Q. and TREICHEL, G. W. (1955). *Ecol. Monogr.*, **25**, 85.

PUCEK, Z. (1960). *Acta theriol.*, **3**, 269.

PUCEK, Z. (1969). *Acta theriol.*, **28**, 403.

REICHSTEIN, H. (1964). *Zeitschr. wissen. Zool.*, **170**, 111.

ROOD, J. P. (1965). *J. Mammal.*, **46**, 426.

ROWE, F. P., TAYLOR, E. J. and CHUDLEY, A. H. J. (1964). *J. Anim. Ecol.*, **33**, 477.

SADLEIR, R. M. F. S. (1965). *J. Anim. Ecol.*, **34**, 331.

SCHNELL, J. H. (1968). *J. Wildl. Managmt*, **32**, 698.

SHEPPE, W. (1972). *J. Mammal.*, **53**, 445.

SHILLITO, J. F. (1963). *Proc. zool. Soc. Lond.*, **140**, 533.

SMITH, M. H., BLESSING, R., CHELTON, J. G., GENTRY, J. B., GOLLEY, F. B. and MCGINNIS, J. T. (1971). *Acta theriol.*, **16**, 105.

SMYTH, M. (1966). *J. Anim. Ecol.*, **35**, 471.

SOUTHERN, H. N. (1964) ed. *The Handbook of British Mammals*, Blackwell, Oxford.

SOUTHERN, H. N. (1970). *J. Zool.*, **162**, 197.

SPENCER, A. W. and STEINHOFF, H. W. (1968). *J. Mammal.*, **49**, 281.

STEIN, G. H. W. (1961). *Z. Säugetierk.*, **26**, 13.

STICKEL, L. F. (1954). *J. Mammal.*, **35**, 1.

STICKEL, L. F. (1960). *J. Mammal.*, **41**, 433.

STODDART, D. M. (1970). *J. Anim. Ecol.*, **39**, 403.

TANTON, M. T. (1965). *J. Anim. Ecol.*, **34**, 1.

TANTON, M. T. (1969). *J. Anim. Ecol.*, **38**, 511.

TARKOWSKI, A. K. (1957). *Ann. Univ. M. Curie–Sklodowska* C, **10**, 177.

TAYLOR, K. D. (1968). *E. Afr. agric. For. J.*, **34**, 66.

THOMAS, O. (1893). *Natur. Sci.*, **2**, 286.

VERHEYEN, W. and VERSCHUREN, J. (1966). *Explor. Parc natn. Garamba Miss. H. de Saeger*, **50**, 1.

WATTS, C. H. S. (1968). *J. Anim. Ecol.*, **37**, 25.

WATTS, C. H. S. (1969). *J. Anim. Ecol.*, **38**, 285.

WOOD, D. H. (1971). *Aust. J. Zool.*, **19**, 371.

ZEJDA, J. (1962). *Zool. listy*, **11**, 309.

ZEJDA, J. (1971). *Zool. listy*, **20**, 229.